EAI/Springer Innovations in Communication and Computing

Series editor
Imrich Chlamtac, European Alliance for Innovation, Gent, Belgium

The impact of information technologies is creating a new world yet not fully understood. The extent and speed of economic, life style and social changes already perceived in everyday life is hard to estimate without understanding the technological driving forces behind it. This series presents contributed volumes featuring the latest research and development in the various information engineering technologies that play a key role in this process.

The range of topics, focusing primarily on communications and computing engineering include, but are not limited to, wireless networks; mobile communication; design and learning; gaming; interaction; e-health and pervasive healthcare; energy management; smart grids; internet of things; cognitive radio networks; computation; cloud computing; ubiquitous connectivity, and in mode general smart living, smart cities, Internet of Things and more. The series publishes a combination of expanded papers selected from hosted and sponsored European Alliance for Innovation (EAI) conferences that present cutting edge, global research as well as provide new perspectives on traditional related engineering fields. This content, complemented with open calls for contribution of book titles and individual chapters, together maintain Springer's and EAI's high standards of academic excellence. The audience for the books consists of researchers, industry professionals, advanced level students as well as practitioners in related fields of activity include information and communication specialists, security experts, economists, urban planners, doctors, and in general representatives in all those walks of life affected ad contributing to the information revolution.

About EAI

EAI is a grassroots member organization initiated through cooperation between businesses, public, private and government organizations to address the global challenges of Europe's future competitiveness and link the European Research community with its counterparts around the globe. EAI reaches out to hundreds of thousands of individual subscribers on all continents and collaborates with an institutional member base including Fortune 500 companies, government organizations, and educational institutions, provide a free research and innovation platform.

Through its open free membership model EAI promotes a new research and innovation culture based on collaboration, connectivity and recognition of excellence by community.

More information about this series at http://www.springer.com/series/15427

Anandakumar Haldorai
Umamaheswari Kandaswamy

Intelligent Spectrum Handovers in Cognitive Radio Networks

Anandakumar Haldorai
Department of Computer Science
and Engineering
Sri Eshwar College of Engineering
Coimbatore, Tamil Nadu, India

Umamaheswari Kandaswamy
Department of Information Technology
PSG College of Technology
Coimbatore, Tamil Nadu, India

ISSN 2522-8595   ISSN 2522-8609   (electronic)
EAI/Springer Innovations in Communication and Computing
ISBN 978-3-030-15418-9   ISBN 978-3-030-15416-5   (eBook)
https://doi.org/10.1007/978-3-030-15416-5

This Springer imprint is published by the registered company Springer Nature Switzerland AG
The registered company address is: Gewerbestrasse 11, 6330 Cham, Switzerland

Preface

The wireless communication sector has certainly been one of the fastest developing sectors of the communications industry in recent years due to the fact that wireless applications have gradually been on increase. The rapid proliferation of wireless technologies is expected to increase the demand for radio spectrum by orders of magnitude over the next decade. This problem must be addressed via technology and regulatory innovations for significant improvements in spectrum efficiency and increased robustness and performance of wireless devices. As a result, various wireless applications and systems operating in unlicensed spectrum bands have gradually led to overcrowding of the spectral bands making them limited and unavailable. However, investigations into the spectrum insufficiency problems by numerous regulatory bodies around the world, including the US Federal Communications Commission (FCC) and the Independent Regulator and Competition Authority in the United Kingdom, have reported that although the demand for spectrum will significantly increase in the near future, the major problem is not the spectrum insufficiency but the inefficiency in spectrum usage.

Emerging cognitive radio technology has been identified as a high-impact disruptive technology innovation that could provide solutions to the radio energy problem and provide a path to scaling wireless systems for the next 25 years. Significant new research is required to address the many technical challenges of cognitive radio networking. An efficient handover decision (HD) mechanism is required to perform switching from one network to another for providing unified and continuous mobile services that include seamless connectivity and ubiquitous service access. The HD involves efficiently combining handover initiation and network selection process. The network selection decision is a challenging task, and it is a central component for making HD for any mobile user in heterogeneous environment that involves the number of static and dynamic parameters. Cognitive radio (CR) is proposed as a robust solution to the problem of inefficient spectrum usage and handover decision-making. As CR can coexist with the existing licensed primary users, efficient protocols are required for performing spectrum sensing and unused spectrum allocation among the secondary users.

This book covers ideas, methods, algorithms, and tools for the in-depth study of performance and reliability of handovers in cognitive radio networks. The field of cognitive radio networks is moving toward a trending research domain by comprising several areas of computer science, electrical, and other engineering disciplines. The scope of *Intelligent Spectrum Handovers in Cognitive Radio Networks* is to explore and contribute numerous research contributions relating to the field of spectrum sensing, network specifications, mobility, energy, and security-aware management system of handover procedures.

In this book, we express the techniques and advances of intelligent spectrum handovers that can be used in overcoming and solving complex tasks in cognitive radio networks. This book is based on various research horizons and contributions focusing on cognitive radio network challenges over:

- Mechanism of cooperative spectrum handovers and its efficiency in cognitive radio networks compared to traditional distributed technique.
- Intelligent cognitive radio communication for rapidly changing network specifications.
- Focus on exploring the energy-efficient spectrum handovers in cognitive network selection.
- Analyzing the efficiency of distributed algorithms for learning in CR networks.
- Explore the basics of software radio architecture for LTE communication.
- Advancements in performing dynamic spectrum handovers.
- Significance of green wireless communications via cognitive handover.
- Utilization of supervised machine learning techniques for dynamic decision-making.
- Methodologies for performing secure distributed spectrum sensing.
- Discover various applications and services of intelligent spectrum handover.

This book opens the door for authors toward current research in cognitive radio network area for future wireless communication systems.

We would like to thank Ms. Mary E. James, Senior Editor, and Mr. Abhishek Ravi Shankar, Project Coordinator (Books), from Springer Nature and EAI/Springer International Publishing AG for their great support.

We anticipate that this book will open new entrance for further research and technology improvements. All the chapters provide a complete overview of intelligent computing and sustainable systems. This book will be handy for academicians, research scholars, and graduate students in engineering discipline.

Coimbatore, Tamil Nadu, India   Anandakumar Haldorai
Coimbatore, Tamil Nadu, India   Umamaheswari Kandaswamy

Contents

About the Authors

Anandakumar Haldorai is Professor (Associate) and Research Head in the Department of Computer Science and Engineering, Sri Eshwar College of Engineering, Coimbatore, Tamil Nadu, India. He received his Master's in Software Engineering from PSG College of Technology, Coimbatore, and PhD in Information and Communication Engineering from PSG College of Technology under Anna University, Chennai. His research areas include Cognitive Radio Networks, Mobile Communications, and Networking Protocols. He has authored more than 85 research papers in reputed international journals and IEEE, Springer Conferences. He has authored six books and many book chapters with reputed publishers such as Springer and IGI. He served as an Editor in Chief of Inderscience *IJISC* and reviewer for IEEE, IET, Springer, Inderscience, and Elsevier journals. He is also the Guest Editor of many journals with Wiley, Springer, Elsevier, Inderscience, etc. He has been the General Chair, Session Chair, and Panelist in several conferences. He is Senior Member of IEEE, MIET, MACM, and EAI research group.

Umamaheswari Kandaswamy Professor and Head, Department of Information Technology, PSG College of Technology, India, has completed her Bachelor's degree in Computer Science and Engineering in 1989 from Bharathidasan University and her Master's in Computer Science and Engineering in 2000 from Bharathiar University. She completed her PhD in Anna University, Chennai, in 2010. She has rich experience in teaching for about 22 years. Her research areas include Classification Techniques in Data Mining, and her other areas of interest are Cognitive Networks, Data Analytics, Information Retrieval, Software Engineering, Theory of Computation, and Compiler Design. She has published more than 100 papers in international and national journals and conferences. She is a life member in ISTE and ACS and fellow member in IE. She is the Editor for *National Journal of Technology*, PSG College of Technology, and Reviewer for many national and international journals.

Chapter 1
Cooperative Spectrum Handovers in Cognitive Radio Networks

1.1 Introduction

1.1.1 Preamble

Wireless communication technology has reached the pinnacle of its entry in the twenty-first century. The recent advancements in Information Communication Technology (ICT) industry wireless communication, in particular, have grown in the areas of wireless data links networks and mobile networks, placing a burden on the scarcity of the wireless spectrum [1]. The pertaining spectrum scarcity has directed recent technologies to emerge in innovating new approaches towards an efficient spectrum usage and management.

The telecommunication regulatory authority—Federal Communication Commission [2] in the United States, and Independent Regulator and Competition Authority in the United Kingdom talk about the increasing demand for spectrum under utilization. This raises the problem of inefficiency in the spectrum usage [3]. The standard regulatory bodies need to take measures to increase the amount of prevailing unlicensed spectrum utilization, in spite of licensing spectrums. However, there are new technologies that use both licensed and unlicensed spectrums, leading to one of the prevailing disputes. Long Term Evolution (LTE) is occurring when an unlicensed spectrum occupies unlicensed frequencies that are operating in parallel with existing licensed wireless standards [4]. This paved the way to a novel approach for spectrum management, to address the spectrum availability and utilization inefficiency. The proposed technique must be capable of providing wireless access to unlicensed cognitive radio users, by allowing them to dynamically gain access towards the unused licensed spectrum. Handover process in cognitive radio is a technique that transmits information from one terminal to another, as the user navigates the network coverage area of the telecommunication system [5]. The

© Springer Nature Switzerland AG 2019
A. Haldorai, U. Kandaswamy, *Intelligent Spectrum Handovers in Cognitive Radio Networks*, EAI/Springer Innovations in Communication and Computing,
https://doi.org/10.1007/978-3-030-15416-5_1

process of handover involves a selection of dynamic networks among diverse network environments, and can be either a mobile or network triggered technique.

Cognitive Radio (CR) technique is applicable in various fields, including public safety, vehicular communications, healthcare, wireless sensor networks, business models, and green computing. These applications need to analyze various issues such as network pricing, communication and data services, standard terminology, network conditions, diverse users and service provider performance [6]. This requires addressing challenges such as energy efficiency, spectrum detection, efficient spectrum allocation and spectrum utilization to evaluate the dynamic handover process in cognitive radio techniques, using supervised machine learning and bio-inspired frameworks for efficient communications [7].

1.1.2 Genesis of Cognitive Radio

The CR network concept proposed by [8] is used to determine the environmental conditions and communication parameters that can be automatically sensed, and need to be customized [9]. The main components of the CR network are primary networks and cognitive networks, which are built over Software Defined Radio (SDR) technology [10]. The primary network consists of licensed or primary users and the cognitive network consists of unlicensed or secondary users [11]. CR technology provides dynamic access of unused spectral bands of licensed users in an efficient way [12].

CR networks are categorized based on the information sharing mechanism, as centralized and distributed networks [13]. The centralized network is also called the cooperative network, where secondary users are managing the CR users. This requires enormous energy consumption to provide better Quality of Service (QoS).

1.2 Overview of the Cognitive Radio

1.2.1 Architecture of Cognitive Radio Networks

CR communicates with external devices, various sensors (e.g., spectrum sensor), and other resources that are accessible through the network and interact with informational resources like web resources. Figure 1.1 represents the overview of the CR and its interactions [14].

Currently, the allocation and utilization of the spectrum are structured and conquered by long planning cycles. Spectrum aware radios can be used for the management of the spectrum into a new structure that is embedded within each individual radio [15]. These radios cooperate to optimize the allocation of the spectrum to meet Radio Frequency (RF) device requirements and a sensor in CR handles these objectives, which are capable of performing spectrum sensing.

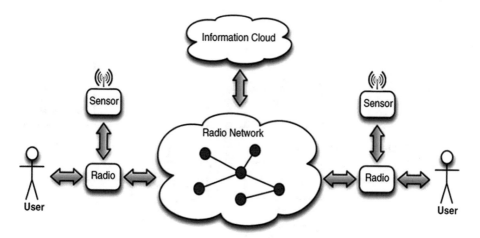

Fig. 1.1 Architecture of cognitive radio networks

CR could provide standardized interfaces to access dynamic networks and support optimization of network resources. Based on the standard protocol stack, CR can be allocated to the particular layers:

Application Layer: This layer includes various communication capabilities and consists of advanced Personal Digital Assistant (PDA) features to support rich functionalities and create interfaces for the user [16].

Network Layer: Allows easy interaction among heterogeneous networks that the radio can reach [17].

Data Link and Physical Layers: Provides better performance like connectivity, bandwidth, and spectrum transmission.

1.2.2 Spectrum Sensing for CR

The area of spectrum sensing has become progressively significant as CR is being used in applications. In many areas, CR systems coexist with other radio systems without causing undue interference to existing systems [18]. When spectrum sensing occupancy is considered, the cognitive radio system must accommodate a range of considerations:

Continuous Spectrum Sensing: It is mandatory for the CR system to constantly sense the spectrum occupancy [19]. Normally PU in a CR network on a non-interference basis and sensing of the spectrum when PU returns during spectrum sensing.

Monitor for Alternative Empty Spectrum: If the PU returns to the spectrum being used, then an alternative space needs to be allocated for existing SU [20].

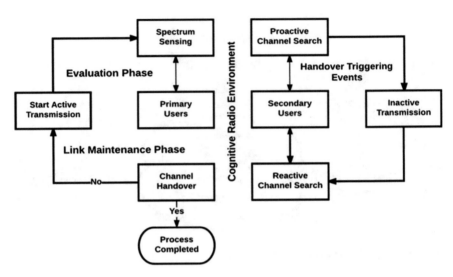

Fig. 1.2 Spectrum handover cyclic process

Monitor Type of Transmission: The CR system should find the PU transmission type and sense the received signal [21]. When this is done, the false transmissions and intrusions are ignored.

1.2.3 Spectrum Handover in CR

CR technology plays a vital role in the spectrum scarcity problem. Spectrum handovers for unlicensed users are performed when licensed users reuse the spectrum due to the spectrum varying nature of CR networks. The spectrum handover can be defined as the process where a SU changes its spectrum space [22]. The challenge during this phase is to find a new available channel and the type of information that is being transmitted.

Spectrum involves two users: first, the appearance of PU in the licensed channel, which essentially makes SU perform handover. Second, spectrum handover can happen due to SU mobility this is when the transmission coverage of the cognitive user overlaps with a licensed user currently using the same channel [23].

Spectrum handover can be explained as a cyclic process with two phases: evaluation phase and link maintenance phase, as shown in Fig. 1.2. During the evaluation phase, CR observes and analyzes whether the handover triggering incidents shall take place or not [24]. When the SU performs handover, it enters link maintenance phase. In this phase, the cognitive user will hand over the channel to the licensed user and maintain data transmission over another available channel. The cycle continues as the SU exits the link maintenance phase.

- **Spectrum Handover Process**: Handover process was mainly triggered due to the arrival of PUs, which found their licensed bands occupied by SUs [25]. In this case, the unlicensed users should instantly initiate the spectrum handover procedure and evacuate the spectrum space for the licensed user.
- **Spectrum Handover Decision Timing**: The process of selecting a target channel for handover can be executed either proactively or reactively according to:

  During priority based handover—the SUs can be served based on their arrival time in order to mitigate the adverse effect of the interruption/handover process and this in turn improves the quality of service, by providing channel utilization to secondary users with a higher priority [26].

  Based on the spectrum environment—the available spectrum bands for wireless communications are categorized as licensed spectrum bands and unlicensed spectrum bands [27].

1.3 Spectrum Sensing Techniques

Sensing a spectrum requires knowledge of creating environmental awareness by enabling the CR to discover a temporary vacant spectrum band, while at the same time detecting the presence or absence of primary users. The process of local sensing is done at the device level, whereas cooperative sensing is performed at the network level, which involves many dynamic users [28].

1.3.1 Cooperative Spectrum Sensing

CR cooperative spectrum sensing occurs when a group or network of CRs distribute the sensed information they gain. This provides a better picture of the spectrum usage over the area where the CRs are located [29]. Cooperative sensing is categorized based on how CR users cooperate and share the sensing data in the network, as centralized and distributed sensing [30].

Centralized cooperative sensing includes a central identity called Fusion Center (FC) that follows a three-step process of cooperative sensing.

Selecting a frequency band for sensing, that instructs all cooperating CR users to individually perform local sensing.

All cooperating CR users report their sensing results via the control channel [31].

The FC combines the received local sensing information, determines the presence of PUs, and disseminates the decision back to cooperating CR users.

Distributed cooperative sensing is a three-stage process that includes local sensing, reporting, and data fusion, which includes fundamental components of cooperative sensing.

Fig. 1.3 Cooperative sensing process

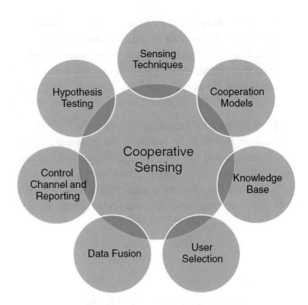

The process of cooperative sensing is analyzed based on the following key elements shown in Fig. 1.3:

- Techniques required for cooperation, called cooperation models
- Various sensing techniques
- Model for controlling channel and reporting
- Data fusion model
- Hypothesis testing
- Selection of user
- Knowledge base

The cooperative spectrum sensing elements are as follows:

Radio Frequency (RF) environment can be sensed via sensing techniques that are used for detecting the PU signal or the available spectrum [32], which analyzes how CR users cooperate with each other.

The presence or absence of a PU can be identified using hypothesis testing, and is a statistical test, performed individually by each cooperating user to make local decisions by the fusion center for cooperative decision.

In control channel and reporting, the sensing result obtained by cooperative data fusion is combined, and the sensing results are shared to make the cooperative decision [33]. The signal combining techniques or decision fusion rules can be used for sensing results based on their data type shown in Fig. 1.3.

User selection is done to maximize the cooperative gain and minimize the cooperation overhead in order to efficiently select the cooperative CR users to find the spectrum space.

Knowledge base includes prior knowledge or accumulated information through experience and may include various details like PU and CR user locations, PU activity models, and Received Signal Strength (RSS) [34].This facilitates the cooperative sensing process to improve detection performance.

1.3.2 Non-Cooperative Spectrum Sensing

Each user determines the presence and absence of PUs independently. This technique is based on detecting weak signals from a primary transmitter. The primary transmitter is based on detection techniques where a cognitive user determines signal strength generated from the PU [35]. In non-cooperative sensing, there is no signaling between the PUs, the cognitive users, and primary receivers are unaware of the cognitive user's location. The non-cooperative sensing is performed by many sensing methods such as:

- Energy detection technique to sense the environment.
- Cyclostationary based sensing which requires some information about the spectral user signal characteristics.
- Matched filter based sensing is a technique that requires the complete information of the spectral user signal.
- Waveform based sensing is a technique which is applicable to systems with known signal patterns.

1.4 Literature Survey

1.4.1 Energy Efficient Network Selection

The [36] analyzed the energy efficiency and QoS performance of 802.11e for low rate applications and compared to 802.15.4 under varying interference and traffic conditions. During energy optimization, 802.11e can achieve higher energy efficiency and QoS. The energy efficiency in 802.11 using distributed coordinated function was analyzed and compared the impacts of various contention windows and packet sizes. It was shown that under error prone environments, the optimal packet size can improve more on energy efficiency than the optimal contention window, and combination of both can achieve the maximum optimization.

The [37] proposed a collaborative spectrum sensing protocol in order to enhance the network energy efficiency by efficient reduction of sensing reports from the secondary users to the fusion center; this method compromised the authentic availability of vacant spectral band. The ability of channel bonding to increase data rate was used to allow more flexibility in distributing the load. These studies rely on the assumption that increasing the channel width should increase the data rate, since more data was being transmitted over a wider bandwidth.

The [38] has proposed a theoretical analysis about energy efficiency in cognitive radio network. It mainly analyzed the physical layer of the OSI model. It gives a guideline of how to standardize the energy management in cognitive networks using secondary users. The sensing time evaluated based on re-occupancy of primary users during the secondary user transmission.

1.4.1.1 Intelligent and Optimized Handovers

In [39] associated Packet Switched Handover (PSHO) to reduce handover interruption because the spectrum sensing was not essential in the handover procedure. In addition, it was easier to have a consensus transmission on their target channel. However, when the spectrum handover is in process, PSHO estimated the number of unavailable channels had increased. Thus, one challenge for the PSHO is to verify the target channels to reduce packet loss probability (overall or individual) and bandwidth fragment ratio.

1.4.2 Supervised Machine Learning Techniques for Cognitive Radio Handovers

The [40] proposed certain investigations that Home Location Register (HLR) should possibly avoid registrations from the previous Visitor Location Register (VLR) during the forward pointer setup. Whenever a user query is transmitted to HLR, the VLR first evaluates the registered user and then follows the sequence of pointers based on current VLR. This approach makes use of the call response and Personal Communications Systems (PCS) mobility model for any users. The above process is useful in identifying the users who receive calls occasionally and make changes in Remote Access (RA).

The [41] proposed a handover management process when it moves from one network access point to another. In this work, the virtual channel reservation was monitored to improve the throughput of the SUs, via minimizing the forced termination and blocking probabilities. A new approach helps the secondary users provoke the spectrum handover and find a proper channel. In his way, the dropping rate of the SUs was reduced and did not reflect on the blocking probability and the throughput of the PU.

1.4.3 Social Aware Cognitive Radio Handovers

In [42] discussed the hypothesis that provides an efficient approach for solving the time variable, multi objective and large-scale problems, discrete binary optimization and continuous nonlinear systems use particle swarm intelligence system. All complicated problems relating to CR such as network resource allocation and dynamic spectrum management are used for enforcing dynamic adaptation based on PSO.

The present CR system using a bio-inspired method. This uses evolutionary intelligence by using particle swarm optimization (PSO) and ant colony optimization (ACO) technique that are inspired by the collective behavior of social insects like swarm, ant, and bees. Thus, this colonial mechanism could accomplish complex tasks which can far exceed the individual capabilities compared to a single insect.

Earlier studies have focused on various issues like energy efficiency, spectrum handover, and machine learning in CR communication systems. However, increasing demands in spectrum standards require research on the cognitive spectrum handover problems and application of intelligent techniques to ensure a dynamic and social CR network.

1.5   Motivation of Cognitive Handovers

Numerous standardization agencies and business networks across the world are functioning on how CR and its enabled applications can be incorporated towards their technology development domain. The standard consent is that CR has a potential to transform new applications, such as dynamic spectrum access (DSA) towards improving the performance, efficiency, and reliability of standard wireless networks.

With the growing wireless communications and network paradigm, efficient and smart utilization of scarce spectral resources has become inevitable. Unlicensed users are allowed to share the spectral resources with the licensed users in a cooperative or distributed fashion. There are still some fundamental issues to be addressed; although, many spectrum sensing and sharing schemes have been proposed to enhance the system performance.

The spectrum handover process is a challenging task in CR network structure, where the CR system should be capable of changing the operating band in order to circumvent the activity of detected PU or by transmitting in the available spectrum hole and enhancing the overall system performance.

CR networks consist of PUs and SUs belonging to multiple networks, operated by different network service providers. The future CR networks are application specific and are either centralized or distributed access networks. Hence various mechanisms for opportunistic SU spectrum access are required in a constantly changing radio environment. The parameters needed for transmission of the CR system must be modified accordingly, based on dynamic bands. It is necessary to propose novel techniques to enhance the sensing performance of CR users and design a DSA scheme that maximizes the throughput of SU.

Hence, the chapter is focused to improve the overall system performance and to investigate different factors that affect the handover performance. A significant handover technique is proposed to improve the performance of the CR network by analyzing the existing state of the art techniques.

1.6 Challenges of Cognitive Radio

The challenges of CR are plenty, such as intelligence distribution and implementation delay/protocol overhead, cross-layer design, security, sensing algorithms, and flexible hardware design. In the following, some of these challenges are identified and described as follows:

Detection, False Alarm, and Miss-Detection Probability
Spectrum under utilization occurs due to false alarms and a missed detection can cause interference with the PUs. In cognitive radio, these issues can violate the rights of the incumbents on the channel, which violates the main principle of CRNs.

Hardware
There are certain constraints in CR wireless sensors like computational energy, storage system and device energy. Designing intelligent hardware for CR is a very challenging issue, since the CR can change the transmitter parameters based on how it interacts with the environment. Dynamic computational intelligence techniques have been proposed to overcome CR basic principles called observation, reconfiguration, and cognition.

Topology Changes
Network topology directly affects the lifetime of CR and deploys applications statically or dynamically. However, it is not always feasible to identify such a path in the static networks topology. An adaptive self-configured topology mechanism is needed, which performs better than static topology in order to obtain scalable, reduced energy and achieving better network performance.

Fault Tolerance
A self-adaptable feature that involves network formation, self-configuration, and auto-healing properties that avoids the faulty node must be derived. The CR should be resistant towards hardware, software, or natural disasters. Designing a fault tolerant network protocol is one of the challenging issues in CR, to ensure that the overall function of the WSNs should not be interrupted.

Manufacturing Costs
The CR applications are currently being deployed in large numbers, resulting in significant reductions in cost. But in contrast to conventional wireless sensor networks, CR requires moderate memory and computational capabilities whose algorithm design requires reduced transmission cost and battery outage which is a demanding issue.

Learning Process

Machine learning has seen tremendous growth in present years where the learning process can be either supervised or unsupervised. It has variable defiance such as (1) avoiding incorrect choices during independent learning processes before a viable decision (2) the objectives and contributions in the context of the CRs need to be defined. The implementation and algorithm designs that are linked to enable devices or networks to learn from past experiences to enhance their behavior are also too complex.

Channel Selection

A channel for data communication in CR needs to be selected using sensors, by negotiating with the neighbors since there is no dedicated channel to send data. There is a lack of cooperation between primary and secondary users since PUs may arrive in the channel and the SUs have to vacate the channel. Here, dynamic AI algorithms are required to intelligently consider the PU's behavior on the channel.

Energy Consumption

CR requires energy for frequent spectrum handoff and the where PUs' activities are sensed via a CR wireless sensor, and there are power constraint devices with less energy. Many applications require diverse antennas for monitoring the PUs' activities in addition to the energy needed for spectrum sensing, thus reducing energy consumption for negotiation of channel, discovering route, and reception of data packets.

Quality of Service (QoS)

Wireless networks need to maintain an adequate level of QoS to avoid harmful consequences in vital applications. The QoS is demanding due to parameters like bandwidth, delay, jitter, reliability and resource constraints like processing power, memory, and power sources in wireless sensor nodes. Since the detection of PUs' channel arrival is difficult, PU's communication should be interference free with the SUs. The primary signal missed detection and false alarm can cause additional disputes.

Security

All recent systems are dynamic in nature and hence, the number of potential interactions led to violation increases. The foremost dispute is on equipment authorization, especially on evaluation standards and security. Software certification and security are also challenging areas when software controls the dynamic system. When the number of combinations of interactions is high and the mobility agility of CRS is great, then the monitoring mechanisms are a challenging task.

1.7   Objective of the Cognitive Handovers

The chapter aims at developing a centralized and cooperative spectrum sensing technique to establish spectrum management, which is capable of optimizing the energy efficiency, spectrum allocation, decision-making, and improving the social communication to attain dynamic handovers in any wireless network. The research has the following objectives:

- To identify energy efficient spectrum sensing technique for cooperative cognitive radio network (CRN)
- To maximize the energy capacity by minimizing the outage probability
- To optimize the sensing time for maximizing the spectrum utilization
- To improve the lifetime of the CRN using active scan spectrum sensing technique
- To analyze energy efficiency of cognitive radio network
- To find a secondary user spectrum allocation
- To perform dynamic handovers
- To use efficient data communication in the cognitive networks

1.8   Cognitive Handovers Contribution

The objectives of this chapter are addressed by proposing cooperative spectrum handovers, to dynamically identify multiple spectrum access opportunities during a single sensing period. This increases the attainable cooperative gain, though reducing the incurred cooperation overhead and sensing errors. Two important aspects of cooperative spectrum sensing are (1) cooperating user selection and (2) reliable fusion of sensing results.

The problem of efficient sensing accuracy has also been addressed and cooperative spectrum sensing through the selection of independent heterogeneous secondary users for parallel cooperative spectrum sensing is performed to make a group of users to sense multiple channels simultaneously. Further performance has been improved by addressing the problem of fusing the local decisions of heterogeneous secondary users by focusing on the reliability of local decisions.

This chapter analyzes the problems in cognitive radio network handovers and the contributions are listed as follows:

- The proposed protocol IEEE 802.16 g is used to perform QoS data streams and improve cognitive capabilities of cognitive radio networks such as (1) minimal battery outage and (2) increased data transmission, thereby ensuring an energy efficient handover.
- The proposed spectrum sensing technique analyzes (1) missed detection probability, (2) false alarm probability, and (3) the probability of sensing error, and hence improving CR spectrum sensing ability.
- A supervised machine learning technique called SpecPSO based on swarm intelligence is proposed for optimizing handovers for performing dynamic handover by adapting to the environment and make smart decisions.
- The QoS parameters have been evaluated for providing handover through (a) effective spectrum usage, and (b) increased data transfer rate at reduced power using proposed Social Cognitive Handover (SCH).

1.9   Performance Measures

The performance of a dynamic spectrum sharing scheme depends on various parameters such as the sensing accuracy, primary user activity, channel bandwidth, and interference. The important measures to evaluate the performance are spectrum utilization, fairness, and throughput.

- **Spectrum utilization**: The percentage of time that the secondary user has access to the channel. If multiple users try to access the same channel, the utilization is divided between them.
- **Fairness**: The performance metric which evaluates whether all the secondary users receive fair share of spectral resources.
- **Throughput**: The rate of successful message delivery over a channel, measured in bits per second. The system throughput is the sum of the data rates of all the users in the system. The maximum achievable throughput is equivalent to the channel capacity of the system.

1.9.1   Energy and Power of Signal

Signal energy and power refers to any random signal $x(t)$ given in Eq. (1.1)

$$E_x = \sum_{n-\infty}^{\infty} |x(t)|^2 \tag{1.1}$$

If the given signal does not decay with respect to time, then the energy will be infinite. Power is defined as the amount of energy consumed per unit time. This quantity is useful if the energy of the signal goes to infinity. If the power consumption at a CR terminal can be reduced, then the energy optimization in the CR networks will be increased.

1.9.2   Signal to Noise Ratio

Signal to Noise Ratio (SNR) is a measure of the sensitivity performance of a receiver to compare the signal and noise levels for a known signal level. The Signal to Noise (S/N) ratio or SNR is normally expressed in decibels.

The greater the difference between signal and unwanted noise, the greater SNR better radio receiver sensitivity performance given in Eq. (1.2).

$$\text{SNR} = \frac{P_{\text{signal}}}{P_{\text{Noise}}} \tag{1.2}$$

1.9.3 Probability of False Alarm

The selection of threshold decision is the right way to separate signal from noise. For any threshold λ, four types of events can occur. They are:

- Target present; $x(t) > \lambda$; correct detection.
- Target present; $x(t) < \lambda$; missed detection.
- Target not present; $x(t) > \lambda$; false alarm.
- Target not present; $x(t) < \lambda$; no action.

The frequency of occurrence of these events can be described using two probabilities.

1. The probability of deciding if a signal present is when there is only noise it is known as the probability of a false alarm, is given in Eq. (1.3)

$$P_{\text{fa}} = \int_{\lambda}^{\infty} P_n(x)dx \tag{1.3}$$

2. The probability of a correct detection is given in Eq. (1.4)

$$P_{\text{d}} = \int_{\lambda}^{\infty} P_{\text{sn}}(x)dx \tag{1.4}$$

The subscripts indicate the PDFs of noise (n) and signal plus noise (sn).

1.9.4 Multimodal Test Functions

Test functions are useful to evaluate characteristics of any optimization algorithm, such as (Table 1.1):

- Convergence rate.
- Precision.
- Robustness.

Table 1.1 Different multimodal test functions

Function name	Objective function	Optimal function value
Sphere	$M \inf(x) = \sum_{k=1}^{n} x_i^2$	30
Griewank	$M \inf(x) = \frac{1}{4000} \sum_{k=1}^{n} x_i^2 - \prod_{k=1}^{n} \cos\left(\frac{x_i}{\sqrt{k}}\right) + 1$	30
Rastrigin	$M \inf(x) = 10n + \sum_{k=1}^{n} \left[\left(x_i^2 - 10\cos(2\pi x_i)\right) \right]$	30
Ackley	$M \inf(\vec{x}) = 20 + e - 20e^{-0.2\sqrt{\frac{\sum_{j-1}^{n} x_j^2}{n}}} - e^{\frac{\sum_{j-1}^{n} \cos(2\pi x_j)}{n}}$	30

1.10 Applications of Cognitive Radio

A more flexible and efficient use of the spectrum in the future opens up exciting opportunities for cognitive radio to enable and support a variety of emerging applications categorized as follows:

In the field of **Military and Public Security** that involves biological, radiological and nuclear attack detection and investigation, certain command, control, battle damage evaluation, intelligence assistant and targeting. The CR enabled equipments play a vital role.

CR can be applied in **Health Care** like Telemedicine and Body Area Network (BAN) utilize **Wireless medical networks** similar to temperature, pressure, blood oxygen, and Electro-Cardio-Gram (ECG). CR plays a critical role in lifesaving applications.

CR can be applied for **Home appliances and indoor applications** such as intelligent buildings, home monitoring systems, factory automation, and personal entertainment. Certain **Bandwidth-intensive applications** like on-demand or live audio/video streaming and tracking its surveillance will use cognitive radio.

Real-time surveillance applications like environmental monitoring, irrigation, vehicle and traffic monitoring, inventory tracking, disaster relief operations, and tunnel monitoring require quick response systems that depend on CR.

Transportation and vehicular networks such as vehicular communications, urban environments, and highway safety devices require CR enabled devices. **Smart grid networks** like emergency report and demand response, and numerous **Cellular networks** such as smart phones, social networks, Long Term Evolution (LTE), and Worldwide Interoperability for Microwave Access (WiMAX) rely on this unique technology to withstand high network and spectrum demands.

1.11   Summary

The spectrum scarcity and under-utilization problem must be addressed via innovative technology, and standard telecommunication regulatory bodies, which need to obtain significant improvements in spectrum efficiency, and thereby increasing the dynamic performance of wireless devices compared to conventional spectrum management methods. Spectrum scarcity remains one of the major issues in the era of wireless communication systems and much of this need to be attributed to inefficient spectrum usage among licensed users. CR is proposed as a robust solution to the problem of inefficient spectrum usage and handover decision-making. As CR can coexist with existing licensed primary users, efficient protocols are required to perform spectrum sensing and unused spectrum allocation among secondary users. The Cooperative Spectrum Sensing (CSS) methodology of CR is implemented and will efficaciously overcome handover challenges such as energy efficiency, spectrum management, and data communication problems.

References

1. Jo, O., Cho, D.H.: Seamless spectrum handover considering differential path-loss in cognitive radio systems. IEEE Commun. Lett. **13**(3), 190–192 (2009)
2. Au, E.K., Cavalcanti, D., Li, G.Y., Caldwell, W., Letaief, K.B.: Advances in standards and test beds for cognitive radio networks: part I [Guest Editorial]. IEEE Commun. Mag. **48**(9), 76–77 (2010)
3. Anandakumar, H., Umamaheswari, K.: An efficient optimized handover in cognitive radio networks using cooperative spectrum sensing. Intell. Autom. Soft Comput. 1–8 (2017)
4. Anandakumar, H., Arulmurugan, R., Onn, C.C.: Computational Intelligence and Sustainable Systems. In: EAI/Springer Innovations in Communication and Computing (2019)
5. Celebi, H., Arslan, H.: Utilization of location information in cognitive wireless networks. IEEE Wirel. Commun. **14**(4), 6–13 (2007)
6. Wang, B., Liu, K.J.R.: Advances in cognitive radio networks: a survey. IEEE J. Selected Topics Signal Process. **5**, 5–23 (2011)
7. Suganya, M., Anandakumar, H.: Handover based spectrum allocation in cognitive radio networks. In: 2013 International Conference on Green Computing, Communication and Conservation of Energy (ICGCE), Chennai, pp. 215–219 (2013)
8. Qing, Z., Sadler, B.M.: A survey of dynamic spectrum access. Signal Process. Mag. IEEE. **24**, 79–89 (2007)
9. Lu, L., Zhou, X., Onunkwo, U., Li, G.Y.: Ten years of research in spectrum sensing and sharing in cognitive radio. EURASIP J. Wirel. Commun. Netw. **2012**, 28 (2012)
10. Wu, Y., Yang, Z.: Coexistence of primary users and secondary users under interference temperature and SINR limit. J. Electron. **26**, 303–311 (2009). (China)
11. Sherman, M., Mody, A.N., Martinez, R., Rodriguez, C., Reddy, R.: IEEE standards supporting cognitive radio and networks, dynamic Spectrum access, and coexistence. IEEE Commun. Mag. **46**(7), 72–79 (2008)
12. Mitola, I.I.I.: Cognitive Radio: an Integrated Agent Architecture for Software Defined Radio. Royal Institute of Technology (KTH), Stockholm (2000)
13. Akyildiz, W.L., Vuran, M., Mohanty, S.: Next generation dynamic spectrum access/cognitive radio wireless networks: a survey. Comput. Netw. **50**(13), 2127–2159 (2006)

14. Clancy, C., Hecker, J., Stuntebeck, E., O'Shea, T.: Applications of machine learning to cognitive radio networks. IEEE Wirel. Commun. **14**(4), 47–52 (2007)
15. Kasabov, N., Zhou, L., Gholami Doborjeh, M., Gholami Doborjeh, Z., Yang, J.: New algorithms for encoding, learning and classification of fMRI data in a spiking neural network architecture: a case on modelling and understanding of dynamic cognitive processes. In: IEEE Transactions on Cognitive and Developmental Systems, vol. 1-1, p. 99 (2016)
16. Azmat, Y., Chen, N.: Analysis of spectrum occupancy using machine learning algorithms. IEEE Trans. Veh. Technol. **65**(9), 6853–6860 (2016)
17. Pratama, M., Zhang, G., Er, M.J., Anavatti, S.: An incremental type-2 meta-cognitive extreme learning machine. IEEE Trans. Cybernetics. **47**(2), 339–353 (2017)
18. Gong, X., Vorobyov, S.A., Tellambura, C.: Optimal bandwidth and power allocation for sum ergodic capacity under fading channels in cognitive radio networks. IEEE Trans. Signal Process. **59**(4), 1814–1826 (2011)
19. Chen, Y.S., Hong, J.S.: A relay-assisted protocol for spectrum mobility and handover in cognitive LTE networks. IEEE Syst. J. **7**(1), 77–91 (2013)
20. Lu, D., Huang, X., Liu, C., Fan, J.: Adaptive power control based spectrum handover for cognitive radio networks. In: 2010 IEEE Wireless Communication and Networking Conference, Sydney, NSW, pp. 1–5 (2010)
21. Thilina, K.M., Choi, K.W., Saquib, N., Hossain, E.: Machine learning techniques for cooperative spectrum sensing in cognitive radio networks. IEEE J. Selected Areas Commun. **31**(11), 2209–2221 (2013)
22. Yuan, W., Leung, H., Cheng, W., Chen, S.: Optimizing voting rule for cooperative spectrum sensing through learning automata. IEEE Trans. Veh. Technol. **60**(7), 3253–3264 (2011)
23. Chen, H., Zhou, M., Xie, L., Wang, K., Li, J.: Joint spectrum sensing and resource allocation scheme in cognitive radio networks with spectrum sensing data falsification attack. IEEE Trans. Veh. Technol. **65**(11), 9181–9191 (2016)
24. Zhang, F., Zhou, X., Cao, X.: Location-oriented evolutionary games for price-elastic spectrum sharing. IEEE Trans. Commun. **64**(9), 3958–3969 (2016)
25. Ma, W., Fang, Y.: A pointer forwarding based local anchoring (POFLA) scheme for wireless networks. IEEE Trans. Veh. Technol. **54**(3), 1135–1146 (2005). https://doi.org/10.1109/TVT.2005.844651
26. Anandakumar, H., Umamaheswari, K.: Energy efficient network selection using 802.16g based GSM technology. J. Comput. Sci. **10**(5), 745–754 (2014)
27. Anandakumar, H., Umamaheswari, K.: Cooperative spectrum handovers in cognitive radio networks. In: EAI/Springer Innovations in Communication and Computing, pp. 47–63 (2018)
28. Anandakumar, H., Umamaheswari, K.: A bio-inspired swarm intelligence technique for social aware cognitive radio handovers. Comput. Elect. Eng. **71**, 925–937 (2018)
29. Gavrilovska, L., Atanasovski, V., Macaluso, I., DaSilva, L.: Learning and reasoning in cognitive radio networks. IEEE Commun. Surv. Tutor. **15**(4), 1761–1777 (2013)
30. Lee, A., Helal, Y.: Sung, Anton, S.: situation-based assess tree for user behavior assessment in persuasive Telehealth. IEEE Trans. Human Mach. Syst. **45**(5), 624–634 (2015)
31. Choi, K.W., Hossain, E., Kim, D.I.: Cooperative spectrum sensing under a random geometric primary user network model. IEEE Trans. Wirel. Commun. **10**(6), 1932–1944 (2011)
32. Haldorai, A., Ramu, A., Murugan, S.: Social aware cognitive radio networks. In: Social Network Analytics for Contemporary Business Organizations, pp. 188–202 (2018)
33. Anandakumar, H., Umamaheswari, K.: Supervised machine learning techniques in cognitive radio networks during cooperative spectrum handovers. Clust. Comput. **20**(2), 1505–1515 (2017)
34. Cabric, D., Mishra, S.M., Brodersen, R.W.: Implementation issues in spectrum sensing for cognitive radios. In: Conference Record of the Thirty-Eighth Asilomar Conference on Signals, Systems and Computers, vol. 1, pp. 772–776 (2004)
35. Yucek, T., Arslan, H.: A survey of spectrum sensing algorithms for cognitive radio applications. IEEE Commun. Surv. Tutor. **11**(1), 116–130 (2009). https://doi.org/10.1109/SURV.2009.090109

36. Shen, J., Jiang, T., Liu, S., Zhang, Z.: Maximum channel throughput via cooperative spectrum sensing in cognitive radio networks. IEEE Trans. Wirel. Commun. **8**(10), 5166–5175 (2009)
37. Zhao, Z., Peng, Z., Zheng, S., Shang, S.: Cognitive radio spectrum allocation using evolutionary algorithms. IEEE Trans. Wirel. Commun. **8**(9), 4421–4425 (2009)
38. Haldorai, A., Ramu, A.: Cognitive social mining applications in data analytics and forensics. Adv. Soc. Netw. Online Commun. **1**, 1–250 (2019)
39. Sadreddini, Z., Güler, E., Çavdar, T.: PSO-optimized instant overbooking framework for cognitive radio networks. In: 2015 38th International Conference on Telecommunications and Signal Processing (TSP), Prague, pp. 49–53 (2015)
40. Wang, G., Guo, C., Feng, S., Feng, C., Wang, S.: A two-stage cooperative spectrum sensing method for energy efficiency improvement in cognitive radio. In: 2013 IEEE 24th Annual International Symposium on Personal, Indoor, and Mobile Radio Communications (PIMRC), London, pp. 876–880 (2013)
41. Xu, H., Zhou, Z.: Cognitive radio decision engine using hybrid binary particle swarm optimization. In: 2013 13th International Symposium on Communications and Information Technologies (ISCIT), Surat Thani, pp. 143–147 (2013)
42. Haldorai, A., Ramu, A., Chow, C.-O.: Editorial: Big Data innovation for sustainable cognitive computing. Mobile Netw. Appl. **1**, 1–250 (2019)

Chapter 2
Intelligent Cognitive Radio Communications: A Detailed Approach

2.1 Introduction

The Cognitive Radio (CR) was exhibited formally by Joseph Mitola in 1999, and since, this idea has been extremely well known with analysts in a few fields, for example, broadcast communications, computerized reasoning, and even logic. Joseph Mitola [1] has characterized the CR as "a radio that utilizes show based thinking to accomplish a predefined level of skill in radio-related areas." Most enquries about CR systems have concentrated on the abuse of unused range as argued by [2]. Be that as it may, the CR hubs have the fundamental characteristics to gain a significant ground in the unwavering quality of the remote systems, which has been less investigated, so that is the reason we were occupied with enhancing the remote connection dependability of a video conferencing application. This article also aims at proposing techniques that will enhance the utility of wireless communication applicable in video conferencing networks from portable terminals considering the application of cognitive radio for CRMT. To further demonstrate this rationale, it is critical to select an appropriate scenario that is effective for the approach given in this paper. The approach is more centered on machine learning, which necessitates further research in enhancing actual-time performance of applications connecting the MAS and the CRMT. This paper initially presents the application and utility of artificial intelligence critical for CR networks. Moreover, the article demonstrates the fundamental application of MAS in CR's context, whereby an explanation of the speculated approach and proactive scenarios are provided.

© Springer Nature Switzerland AG 2019
A. Haldorai, U. Kandaswamy, *Intelligent Spectrum Handovers in Cognitive Radio Networks*, EAI/Springer Innovations in Communication and Computing, https://doi.org/10.1007/978-3-030-15416-5_2

2.2 Artificial Intelligence and Cognitive Radio

Artificial intelligence (AI) methods for basic learning techniques and decision application can be connected to structure proficient cognitive radio frameworks. The idea of machine learning was connected to CR for limit amplification and dynamic range. Distinctive learning calculations can be utilized in CR systems (Hidden Markov Model, neural systems, hereditary calculations, choice trees, and fluffy rationale or characterization calculations). Cognitive radio networks should be capable of adapting and learning their wireless connection in reference to an ambient radio ecological framework they are in. An intelligent algorithm like the one centered on machine learning, fuzzy controls, and genetic algorithms are fundamental due to the application of cognitive radio initiatives. Generally, these algorithms are applicable for observation of the actual status of wireless ecological frameworks and enhancing the overall knowledge related to ecology. This information is utilized by a cognitive radio to adjust its choice on range get to. For instance, an auxiliary client can watch the transmission action of essential clients on various channels. This empowers the CR to fabricate learning about the essential clients' movement on each channel. The CR to choose which channel to get to with the goal that the coveted execution destinations can be accomplished then utilizes this information (e.g., throughput is augmented while the obstruction or impact caused to the essential clients is kept up underneath the objective level).

In a wide sense, the thought behind this progressive and innovative proposition comprises of making cognitive radio gadgets to have the capacity to convey effectively, when a significantly trustable identifying approach has guaranteed the nearness of range openings. In a dimensional accumulation, these are the fundamental materials, applicable in twofold: longitude, repeat, time, and scope. Moreover, these include top impedance to others, alongside the compulsory arrival of the radio assets when a client having higher order has returned to the medium. With the end goal to do as such, a CoRa handset needs to detect its condition, to comprehend it, to remove ends, and to find out about it. Moreover, it is obliged to find its potential chances, to settle on choices in regard to its versatility and configuration to validate its QoS whereas keeping note of any forms of transitions in RF intermediate.

Since every one of the suggestions expressed over, this new age of radio gadgets is considered to have discernment capacities, being the purported range detecting assignment presumably the most essential, complex, and testing system. Similarly, a decade has elapsed since CR was accustomed with the obligation to broadcast communication sectors. In the present days, fundamental measures of logic formulations are progressively aiming at launching strategic methodologies of detecting CR settings [3]. Notwithstanding, it could be surmised that numerous years (perhaps over 10 years) are still to come until the point that we can install useful answers for every one of the difficulties associated with the task of a completely CR. The founding stone to this literature study is defined from data analyses in the present that are innately being obtained at the innovative essential systems. Along these lines, the cognitive radio correspondences would be executed under a helped discovery plot given by the essential destinations (i.e., base stations, spine). Based on these lines, the fundamental cornerstone analyzed would be applicable in the advantage of formulating CR innovations.

2.3   Principles of Mobile Radio Communications

In mobile communication system, the radio channels introduce randomness to the signal at the receiver as a result of the presence of two types of fading, which have been classified as large-scale fading and small-scale fading.

2.3.1   Mobile Radio Communication Principles

Radio networks, in mobile communication frameworks, demonstrate a degree of randomness that signals the receiver due to the availability of two critical forms of fading: large-scale and small-scale Fading.

2.3.2   Large-Scale Fading

The substantial scale blurring manages spread models that gauge or anticipate the normal flag control over various transmitters and beneficiaries, more than a thousand meter that definitely ends being applied when analyzing the entire transmitter sequence. This type of fading considers for the unmistakable common and fabricated arrangements (e.g., slopes, woodlands, gatherings of structures) which are regularly found in the landscape that protects a remote correspondence connect. In literature, it is conceivable by [4] to discover a few models managing an assortment of utilization and in addition natural designs for evaluating the mean-way misfortune because of huge scale blurring.

2.3.3   Concentrated Fading

The concentrate form of fading is rotated over different models with frameworks that are an outline of normal transition that happen in a standardized form of ampleness, i.e., the 30 dB/40 bB interest. This ampleness is denoted over a considerable period of travel duration, i.e., the half the overall wavelength or timeframe. Exactly the moment when elements that placate the augmented structure like diffraction, spreading, and reflection are fundamental in sequence, there will be no differentiable form of pathway. The fundamentality of the eliminated pennants in stipulated confinements of CR carries on as administered by a Rayleigh likelihood thickness work. On the other hand, if there is a dominating part between the distinctive arriving courses and the little scale, then obscuring is depicted by a Rician's probability thickness assumption.

2.3.4 Small-Scale Fading

The firm definition of the proposition is related to the small-scale fading illustration referred to as the deferral expansibility of time-transited framework. The illustrative framework of the spread flag indicates the varied controls, which are an element of timing delay. When this illustration is evident, it is concluded that the radio framework undergoes a recurrence of particular blurring, if the most extreme overabundance postpone Tm, it surpasses the image time Ts (i.e., Tm > Ts), which produces channel-prompted between image impedance. Though if the contrary circumstance happens (i.e., Tm < Ts), the channel would experience level or recurrence non-specific blurring [5]. Thus, a practically equivalent to characterization of the flag scattering can be seen in the recurrence space, ordering the channel as recurrence specific blurring if the intelligibility transmission capacity f0 is littler than the flag data transfer capacity W (i.e., f0 < W ≈ 1/Ts), implying that f0 to be a little recurrence run contrasted and the involved by W. On the other hand, if f0 > W ≈ 1/Ts, at that point it tends to be accepted that all the flag's unearthly segments will be influenced likewise the framework (through fading or non-fading).

The duration of fluctuation of the framework: This illustration is formulated due its feature of multiple sections that are entirely dependent to the flag's recurrence and on the situation of the reception apparatus. Nevertheless, due to relative movement between the transmitter, recipient, and similar time-variation conduct of the framework, it can be stimulated by the enhancement of encompassed items.

In reference to time-space, the obscuring speed that a channel might undergo refers to moderate obscuring when soundness timeframe "To" surpasses a picture period allotment Ts (that is To >Ts), reasoned that the channel will proceed as before in the midst of a couple or perhaps one transmitted picture. In situation (i.e., To <Ts), the fast obscuring happens to provoke a couple of exceptional transitions stipulated at the framework when the picture transforms under the stimulation of wavelength twist resilient unalterable slip-up degrees. As for a repeat space, its moderate obscuring takes place if banner information exchange limit is more significant than the horrible extending before the "Doppler" extent (that is W > fd or about 1/Ts > 1/To). Exactly once "W" < fd, the differentiable framework is denoted as the prompt obscuring.

Table 2.1 Examination challenge of the proposition

Serial number	Relative time (µs)	Average relative power(dB)
1	0.0	−4.0
2	0.2	−3.0
3	0.4	0.0
4	0.6	−2.0
5	0.8	−3.0
6	1.2	−5.0
7	1.4	−7.0
8	1.8	−5.0
9	2.4	−6.0
10	3.0	−9.0
11	3.2	−11.0
12	5.0	−10.0

2.3.5 The LTE Link and the Versatile Framework

At the point when the study of intense obscuring parts had been concluded, LTE's overall consideration and adaptable framework of the SC-FDMA channel was fragmented because they all pushed towards examination opposing this suggestion.

2.3.6 The LTE Portal Channel

In reference to the 3GPP specialized details, 45.005 V9.3.0 (2010–05), a few spread structures (e.g., country zone, sloping landscape) are characterized regarding a course of action frameworks, whereby all were characterized considering a certain delay moment with typical control [6]. Emphatically, a framework portraying a normal instance of borough locale is picked representing an examination challenge of the proposition, as indicated in Table 2.1.

As far as the little scale obscuring instruments are concerned, the particulars have showed up in Table 2.1 indicated above, which displays the profile of energy delay, implying a timeframe of flag spread. In that regard, including the expecting relations of work from 0.5, the soundness exchange speed that represents the framework is generally 199 kHz and a while later, the channel can theoretically be assigned repeat particular in reference to base LTE's framework and information transmission identical as 1.5 MHz that doubtlessly outperforms f0. Resultantly, it is termed as driving the length of the variables prefixes (that is customary and expanded) planning a timetable opening of 0.5 microseconds is consolidated into the LTE. Furthermore, going for an addition of the period moving definition onto a redirect portrayed in the

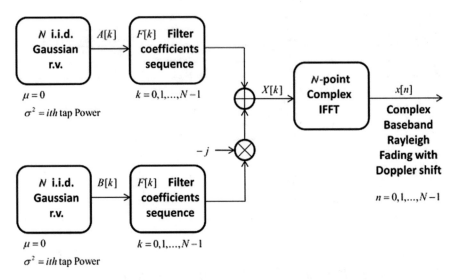

Fig. 2.1 Doppler Rayleigh Fading

above Table 2.1, the time related Rayleigh observed contingency upon a repeat
Doppler, and its square chart is shown in Fig. 2.1.

The system showed in the figure above is an upgrade (i.e., to the extent reminis-
cence possessions with the amount assignments) committed computation at first
shown by Smith for making related Rayleigh self-assertive variables. A strategy
begins with a delivery of data in dual progressions made out from free in addition to
unclearly scattered Gaussian's self-assertive determiners. The variables are
represented by a minimal mean, including a recommendable change in reference
to typical power, which addresses, nonexistent section of the lone rap (i.e., a
comparative framework must be reiterated for all of the taps). At that time, applying
effective channel constants Fk (i.e., according to the desired most extraordinary
Doppler repeat fd, carrier repeat fc, with the "UE" and V), progressions Ak with Bk
are all determined. These are respected preceding the quadrature Xk that is changed
based on timeframe region xn with the true objective to give the necessitated
baseline Rayleigh fading considering Doppler's move.

As of now towards the completion the concluding part, it was made reference to,
which soundness timeframe chooses a typical proportion in midst when the framework's feedback is regarded as the invariable [7]. This indicates that there is an
average general command on enlisting clarity timeframe representing a segment that
defines the Doppler's spreading (fd = V divided by λ) denoted by:

$$T_0 \sqrt{\frac{9}{16 \times \text{fd}^2}} \tag{2.1}$$

The articulation showed up above works out as intended because of including a
geometrical formula that incorporates fundamental essentialness, including an

acumen timeframe ($T_o \approx 1.0/fd$), whereas the description, as recognized by channel's affiliation to seems to be more undeniable 1.0. In the event that we make utilization of the above standard considering a client hardware working on approximate 1800 MHz linking the second working band that operates on a velocity proportionate to an approximate of 15 km per hour, which meant that, we shall acquire $T_o = 16.462$ ms.

Likewise, in case we imply the LTE's methodology, it is expected that considering specific situations (including in the event that we consider for instance an ordinary cyclic prefix) with the end goal to have seven images for every schedule opening at that point around 230 SC-FDMA images would encounter similar conditions like nulls, pinnacles, or something in the middle of until the point that the channel changes once more. At the transmitter, the system starts in the time area with the sequentially organized images, which are regulated, afterward; the assumptions are argued to be parallel and according to an N point Fourier transition, that incorporates the format for adjusting standardized numerals in a repeating framework. Currently, the stipulated data is supplied to the N carriers like the map framework. This happens considering a vital aim of outlining the transformed numerals in a sequence planning of "Coterminous carriers whereby "0" is denoted as M and N are unutilized carriers.

Not long when the M points backward, "Fourier" change will conform to the change of data on timeframe zone, that changed over onto progressive shape including recurring preface, letting consequently the baseline SC-FDMA pictures orchestrated and transferred using a framework. The roundabout intricacy in the middle of SC-FDMA picture including its framework occurred in game plan at starting strides of beneficiary contains clearing the cyclic prefix. These are also defined as changing over the data on a corresponding shape using "M" facilitates "FFT" towards changing the received data to the recurrent space; afterward, the mapping system undergoes totally guarding the N subcarriers. These were appropriated to the client being insinuated, which are offset with a definitive target to give a dominating estimation of the at first balanced composite amounts, which change into timeframe-district to obtain information that a demodulator needs, this gives a recuperated picture that last long in progressive packaging.

Researchers associate correspondence by manhandling the little scale clouding structures for LTE's helpful framework. Considering this juncture, we are extremely certain based on the suggestion considering timeframe-fluctuating channel's definition, including recurrent discernment, aimed at conveying an assistant correspondence into the SC-FDMA structure. Eventually, the running with segments depends on clearing up in detail how this model limits.

The overlay auxiliary correspondence is shown in beginning stage, the recurrence area portrayal of a regular transmission is given as follows:

$$Y(k) = H(k)X(k) + Z(k), \tag{2.2}$$

$Y(k)$ implies the received banner on the kth subcarrier, H_k current advantage, X_k conveyed banner, Z_k WGN test. Of course, Y is without any other individual the whole SC-FDMA gotten picture consolidating a plan of carrier x that had been deformed from H framework, and what is more by the AWGN uproar Z. In the above segment, an equalizer is evident in a getting-together restraint going for reimbursing the mutilations conveyed in its framework. During the examination, an exceptional MSE conveyor had been seen as introduced in a receptor that gives the increment element to kth's carrier denoted by Eq. (2.3) indicated below:

$$G(k) = \frac{H(k)^*}{H(k)^* + \sigma^2} \tag{2.3}$$

whereby * implies kth's conjugate course advantage, in addition to σ^2 and WGN commotion contrast. Something basic being highlighted in this section is MSE stabilizer, which necessitates channel learning aimed at fitting action, and data demonstrated by 3GPP LTE denoted as eNode B in each transformation timeframe. The consideration of an obtained carrier in MSE follows up the gotten flag as shown in Eq. (2.4) below:

$$X(k) = \frac{H(k)^2}{H(k)^2 + \sigma^2} X(k) + \frac{H(k)^*}{H(k)^2 + \sigma^2} Z(k) \tag{2.4}$$

The left part of Eq. (2.4) above shows the flag, while the right part indicates the clamor. In that regard, the flag to commotion that indicates a proportion at it beneficiary is denoted by:

$$\text{SNR}_{\text{MMSE}}(k) = \frac{P\left(\frac{H(k)^2}{H(k)^2 + \sigma^2}\right)2}{\sigma^2\left(\frac{H(k)}{H(k)^2 + \sigma^2}\right)2} = \frac{PH(k)^2}{\sigma^2} \tag{2.5}$$

P insinuates the approximate energy transferred to the flag. The details include banner hullabaloo extent particularly connected to its network settings secured to the kth's carrier. At a point when the standard program is delineated, psychological radio's featured model is in correspondence and overlapped at its fundamental broadcast depicted into much detail. Consistent with the recommendation, significant fading and its frameworks zeniths shall accept an activity considering an incredible frequency settings that will be trailed onto a psychological wireless client, this shall be allowed to be transmitted completely (that is basic assets pretentious with, for instance, unprecedented settings).

As demonstrated by the suggestion, significant obscuring and the channel tops (that is incredible settings) shall accept an activity carried out by engaging operators aimed at working up keen discretionary correspondences by the psychological radio customer. A UE on the LTE system is encountred based on particular carrier numeral, like transistor assets is dedicated to intellectual wireless contraption aimed at working up a sagacious correspondence that will be comparable. After that, Eq. (2.5) is balanced once an intellectual receiver correspondence overlays a standard conduction.

$$Y(k) = H(k) + Z(k) + H_{SU}(k)X_{SU}(k), \quad k \in k, \tag{2.6}$$

However, H_{SU} and X_{SU} demonstrate the channel and the exchanged information identifying with SU separately. Furthermore, the k underscores a simple subcarrier, which encounters any invalid otherwise an apex (that is dependent upon an examination) that shall be reused in SU. The last infers the setting's bit exceeding making impedance over the conformist transfer, which shall simply impact the foreordained subcarrier's number (that is less to the overall count of available carriers) which may vanish if there is no negative settings in the transistors' association trailed by the SU. Thusly, considering the assumption, the assessed transmitted picture consequent to using the MMSE equalizer is achievable considering Eq. (2.7) below:

$$X(k) = \frac{H(k)^2}{H(k)^2 + \sigma^2}X(k) + \frac{H(k)^*}{H(k)^2 + \sigma^2}Z(k) + \frac{H(k)^*H_{SU}(k)}{H(k)^2 + \sigma^2}X_{SU}(k), \tag{2.7}$$

In a manner to which the elements indicated above conform to the flag, the mode of commotion and obstruction are regarded separately. Elements that make up Eq. (2.7) are conceivable and portray the flag's SIR on kth subcarriers as indicated in Eq. (2.8) below:

$$\text{SIR}_{MMSE}(k) = \frac{PH(k)^2}{P_{SU}H_{SU}(k)^2}, \tag{2.8}$$

where PSU alludes to the dispatched power by the forwarded optional client. Then again, the flag to commotion in addition to obstruction proportion can be changed as shown in Eq. (2.9) below:

$$\text{SIR}_{MMSE}(k) = P\left(\frac{H(k)^2}{\sigma^2 + P_{SU}H_{SU}(k)^2}\right), \tag{2.9}$$

Adding to other factors, Eq. (2.9) shown above permits the assessment of CR effects and its correspondence on an ordinary synchromesh.

2.3.7 Neural Networks

A multilayered neural system was utilized to model and gauge the exhibitions of IEEE 802.11 systems. This neural system gives a discovery model to the non-direct connection between the information sources and the yields. This neural system model can gain from preparing information, which can be gotten in an online way when the ongoing estimation information is accessible. In this manner, this model is reasonable for a cognitive radio system for which an incite reaction to the changing radio condition is required from an unlicensed client.

2.3.8   Fluffy Logic

The fluffy rationale is frequently joined with neural systems that can adjust to the earth amid the development of a CR framework. A fluffy rationale control framework can be utilized to get the answer for an issue given loose, loud, and fragmented info data. So, rather than utilizing entangled numerical plans, fluffy rationale utilizes human-reasonable fluffy sets and deduction principles to acquire the arrangement that fulfills the coveted framework targets. The principal preferred standpoint of fluffy rationale is its low unpredictability. In this manner, the fluffy rationale is appropriate for continuous cognitive radio applications in which the reaction time is basic to framework execution. When all is said in done, there are three noteworthy parts in a fluffy rationale control framework: fuzzifier, fluffy rationale processor, and defuzzifier. While the fuzzifier is utilized to delineate fresh contributions to fluffy sets, the fluffy rationale processor actualizes a deduction motor to acquire the arrangement dependent on predefined sets of standards. At that point, the defuzzifier is connected to change the answer for the fresh yield.

2.4   Multi-Agent Systems and Cognitive Radio

The relationship of MAS and cognitive radio may give an extraordinary future to the ideal administration of frequencies (in consideration with the unbending control strategies proposed by the media communications administrators). Due to utilization of unlicensed groups, the CR terminals need to organize and collaborate to best utilize the range without causing impedance. The creators propose a design dependent on specialists where every CR terminal is furnished with a perceptive operator; there are modules to gather data about the radio condition and obviously, the data gathered will be put away in a mutual information base that will be gotten to by all specialists [8]. The proposed methodology depends on helpful MAS (the specialists have regular interests). They work by sharing their insight to expand their group and individual gain.

Operators are conveyed on the PUs and SUs terminals and collaborate with one another in progress proposed in. By agreeable MAS, we imply that PU specialists traded tuples of messages with the end goal to enhance themselves and the area of SU operators. They suggest that the SUs should settle on their choice dependent on the measure of accessible range when they locate an appropriate offer (without sitting tight for reaction from all PUs). As it were, the SU specialist ought to send messages to the proper neighbor PU operator and obviously the concerned PU must react to these operators to a concurrence on sharing the range. After the finish of the range utilization, the SU must pay the PU. An examination is made in the middle of a specialist and a CR [2]. Fundamentally, the two are known about their encompassing surroundings through collaborations, detecting, observing and they have self-sufficiency and authority over their activities and states. They can fathom the relegated assignments freely dependent on their individual abilities or can work with their neighbors by having continuous data trades. To make the CR frameworks functional, it necessitates that few CR systems exist together with one another. Nevertheless, this can cause obstruction [8]. The creators feel that to cure this issue, the SU can coordinate to detect the range and to share it without making obstruction of the PU. For this, they propose plans to shield the PU from impedances by controlling the transmission intensity of the cognitive terminal.

The creators propose collaboration among PUs and SUs and between SUs as it were. Operators are sent on the client's terminals to participate and result in contracts overseeing range assignment. SU specialists exist together and coordinate with the PU operators in an Ad hoc CR condition utilizing messages and instruments for basic leadership. Since practices of specialists are agreeable and caring; it empowers them to amplify the utility capacity of different operators without including costs result as far as traded messages. Nevertheless, the distribution of assets is a critical issue in CR frameworks. It may be finished by arranging among SUs. Specialists propose a model dependent on operators for the range exchanging a CR framework. In any case, rather than arranging range specifically with the PU and SU, a repre-sentative specialist is incorporated. This implies the hardware of PU or SU does not require much insight as it does not have to play out the range detecting. The target of this exchanging is to boost the advantages and benefits of specialists to fulfill the SU. The creators proposed two circumstances, the primary uses a solitary specialist who will misuse and command the system, in either case there will be a few contending operators.

The creators concentrate in the utilization of CR in remote LANs and the likelihood of presenting the innovation of specialists; at the end of the day they attempt to tackle the issue of radio assets distribution by consolidating assets administration WLAN in a decentralized situation, by utilizing MAS. For this, they propose a methodology dependent on MAS for sharing data and choices dispersion among different WLANs in a disseminated way. Scientists utilizing a CR to deal with the handover [6] can decrease impedance from the procurement of the directors in a phone framework amid handovers. To be sure, the portability of the gadget forces an alternate conduct while evolving zones. The terminal must guar-antee benefit coherence of uses and the viable range administration. The creators propose a methodology that utilizes transaction, getting the hang of, thinking and expectation to know the requirements of new administrations in present day remote systems.

In this chapter, the writer points out an algorithm that should be implemented using versatile terminal amid the subjective period used during handover. Thus, MAS comprises few intellectual operators which interrelate with one another. Every agent has the ability of detecting and studying. Additionally, the operator chooses practices dependent on nearby data with endeavors to expand in system execution processes. According to research, another methodology dependent on multi-specialist support realization that is utilized within the CR systems with impromptu-decentralized power. In the end, brand new CR situations would be drafted which would significantly influence prospective case as a return or punishment. In the long run, the aftereffects of the stated methodology exhibit that having the technique at hand will lead to convergence of these systems in compliance to the stated range allotting, which in the end reduces the impediments using the PUs. An exceptionally intriguing methodology anticipated, explaining how the writer has linked learning RL for multiple agents (MARL) and one agent (SARL) in order in accomplishing compassion and aptitude. Thus based on its outcomes it is shown that the MARL and SARL play a joint activity which provides excellent implementation over the entire system.

At long last it was stated that supporting of educative algorithm is adjusted in order to be connected with various application mappings. Therefore it is stipulated that it involves the use of a learning system since the common MARL is always availed to a single agent. Confined learning gives a reward to every specialist so it can settle on the correct choice and pick the best activity. The SU nodes act as learning operators as the receiver and transmitter share typical knowledge and learning results. Here the scholar introduced a LCPP system (Locally Confined Payoff Propagation) that has an imperative capacity for the fortification of learning based on MAS in order to accomplish optimal results through the collaboration of agents within the circulated CR systems. In this case, there will be determination of a channel selection process for the multi-channel and the multi-user as shown [9]. Thus in order to stay away from impact brought about through non-coordination, experience-based channels should be chosen with the use of the SU. MARL system is linked concerning the Q-learning by thinking about the SUs as a factor within the environment. Centered on the stated plan, every SU detects channels after that it chooses a low frequency channel to convey the information, as though no SU user exists. On the off chance that two SUs pick a similar channel for data broadcast, it will result in collision of another and the information package will not be decoded by the recipient. In any case, the SUs can endeavor to figure out how to maintain a strategic distance from one another [10].

According to research by [8], the writer utilized the MAS to plan another discernment cycle using a complex collaboration between SU, PU, and remote conditions, and they utilized the shrouded Markov chains to exhibit the associations between the environment and the user [11]. Thus the impact of the proposed methodology has demonstrated that this algorithm ensure decency among clients. The presence of a reenactment structure to test the proposed works and methodologies can make the utilization of MAS in the CR intriguing more concrete. It is absolutely what the author proposed. Their stage permits considering the developing perspective, and the practices of heterogeneous CR systems [12].

2.5   Service Quality Employed for Video Conferencing

After rapid development of innovative administrations, for instance, videocassette streaming alongside video conferencing, there has been increased necessity of considering each framework one by one and realizing how to distinguish services that ends up primordial [13]. An intuitive video conferencing cannot endure long impediments since there is not sufficient time for the retransmission of the lost packages. The misplaced packages or late ones are essentially overlooked which results in weakening of the pictures and sound. With a system that offers a satisfactory throughput, there is need for managing deferral (amid the transmission and reception of the package) and jitter (setback difference); as a rule this sort of submission cannot endure substantial jitter which might interfere with the sound and picture, thus obviously we should likewise control the loss of these pictures [14].

In any case, we feel that the CR nodes have vital characteristics that guarantee the association progression of video conferencing and in this way assure a decent nature of service administration [15]. In the writing, we discovered that having a decent QoS during video conferencing is vital due to the stated reasons below:

- Packet loss must be $<1\%$.
- Jitter must be <30 ms.
- Delay must be <200 ms.
- Throughput must be >384 Kb/s.

Notwithstanding the lack of genuine information utilized within the CR systems, which relates to the entire network, it is necessary to presume the position of an expert to dole out the required information used within the simulations [16].

2.5.1   Single CRMT Problems and Prospective Solutions

Figure 2.2 shown below demonstrates a way pursued by mobile subscribers when changing to a territory when indication eminence diminishes within an unsuitable position (exhibited by color red) because of range inclusion; it is recommended for all customer to utilize film conferencing rather than other course.

2.5.2   The Anticipated Resolution

Subsequent to numerous occurrences, the CR ought to know about the issue. At that point, various geo-locations or else the capacity of taking note of time of day an incident occurs, radios are in a position of predicting the distinction covered and thus notice the fundamental signal of the main station which will change the quality of signals when clients get to a deficient network coverage [17].

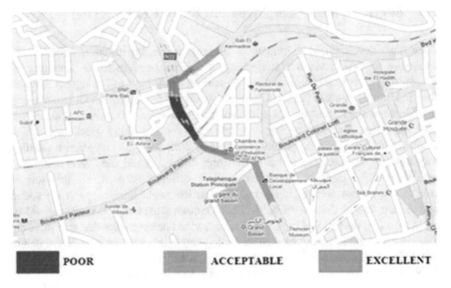

POOR ACCEPTABLE EXCELLENT

Fig. 2.2 Cognitive radio wave eminence

2.5.3 Application

Considering the above statements, video conferencing is utilized in account of a versatile mobile client in need of taking the course pathway wherever the pointer value diminishes further up to an unsuitable phase because of wide range exposure, thus providing an absolute QoS that is low [18]. On the other hand, all thus can be aided by the use of the CR, yet risks still emerge: WHEN cognitive radios used and WHY?

Nearly all researchers have pointed out that the QoS used in video conferencing is thus considered as a throughput using an appropriate parameter. Consequently, the "Throughput" is selected as a solitary appropriate parameter for the function. As a result, the classification of the throughput is necessary, and as we assume the position of a certified professional, we have made our own database following certain principles with the end goal to apply our technique. There was isolation of the database into two major divisions; they included learning segment and the testing segment [19]. It was estimated that there will be transition of the throughput values amid the day even on a similar course; therefore, our actions have been considered for about 1 month and 1 week at three unique interims which are at eight to eleven, eleven to three, and three to five in the evening barring at the end of the week.

The throughput values were purposed to be influenced using three classes:

- Bronze test when a throughput value is below 160 Kb/s. Thus it implies video conferencing is not acceptable, with the class intriguing us since this is the point at which the CR is utilized.

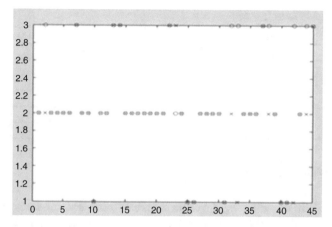

Fig. 2.3 The Grouping A Neural Networking System

- Silver tests considering the throughput ranging from 160 to 384 Kb/s, of satis-factory value.
- Gold throughput test is more prominent compared to 384 Kb/s, which assures absolute contentment of client.

Question one "WHEN"? In order to characterize information, three distinct algorithms are utilized and acquired from the learning machine [20].

- C4.5 calculation of the decision trees
- A multilayer perceptron calculation (neural systems of MLP)
- k-closest neighbor's calculation (K-NN) that comprises a regulated characterization calculation

It can be proved that with a multilayer perceptron not providing acceptable outcomes in contrast to other alternate calculations; 100% value cannot be achieved, regardless of the difference in parameters, for example, the quantity of concealed layers, the quantity of periods lastly the learning pace shown in Fig. 2.3.

In this chapter, it can be commented likewise that the two different calculations which are C4.5 and KNN provide almost 100% result. Although investigating on the created tree, it is viewed that it is not actually, what we need because his GOLD range begins at a rate of about 380 Kb/s rather than 384 Kb/s, which in the end creates errors through misrouting a few occurrences. Relating to the K-NN calculation, the K values were tested using the test database and in most occasion dissimilar outcomes were unique, however it is palatable upon $K = 6$ [21].

In our methodology, the calculation was selected (Table 2.2), since are provided excellent outcomes (demonstrated as follows) regarding unwavering quality and its clustering.

Considering $K = 1$: Throughput models were clearly aligned within the suitable division. Though when $K = 2$, there was misrouting of the data.

Table 2.2 Use of KNN in classifying data

Values of K	Instances correctly classified		Misclassified instances	
$K = 1$	20	100%	0	0%
$K = 2$	19	95%	1	5%
$K = 3$	19	95%	1	5%
$K = 4$	18	90%	2	10%
$K = 5$	18	90%	2	10%
$K = 6$	18	90%	2	10%

2.5.4   Account Description

As indicated by the grouping data, the cognitive radio transmissions are enacted every day from eight to eleven and three to five in the evening in light of the fact that the interims throughput values will be categorized within Bronze class, which is less than 160 Kb/s. Thus for broad guidelines, it is intriguing to think about alternate QoS video conferencing restriction through the arrangement of these information utilizing different techniques used for man-made reasoning, for example, fluffy rationale and hereditary calculations.

Presently, the importance of the CR will be legitimized by assuming the range detection is as of now beneficiary from the versatile unit aimed to be used in this situation as a multimode remote correspondence unit (the MWCT), which is in a position of harboring numerous entrance advancements, for example, the UMTS, GSM, and the WiMAX. Taking into account that the range is not utilized absolutely, the frequency bands can be evidenced by grouping into two divisions such as primary set comprising the free ones and second set consisting of the occupied bands. Additionally, the versatile terminal should change to a fallow recurrence band inclusive of the ones found within the free band group. This methodology involved the study of various situations which concentrated in demonstrating the CR value which is dependent on the quantity of open bands together with the period at which each is active, thus it was identified that there were three conceivable situations which include:

- Adverse (viewing negatively): In this case, the receiver cannot point out an open band (void set), otherwise it distinguishes a few groups; however, its utilization meddles with the primary clients. For this situation, the CR is not utilized in light of the fact that the secondary client ought not to irritate the common clients.
- Constructive (preeminent case): a receiver identifies open band and utilizes it amid the distance with no primary user's intrusion.
- A common N recurrence that has N jumps: in this section, the terminal employs the use of open band (b1), at that point intrusion is stimulated by a primary client, therefore it changes to another open band (b2).
- In the event that the primary client b2 requires his band, then the secondary client should change once more, et cetera, in anticipation of the point when he settles on the CR (prior the return to its previous band frequency), he will thus experience N bounces in return.

Based from our application, the numbers of jumps are dependent on the CR amid the distance between the secondary clients. Every situation specified in this case was computed based on the time of intrusion that was the essential time used at the terminal in acquiring an open band and consequently utilizing it. Thus the intrusion time is characterized as shown below:

$$\text{Time intrusion} = (T\text{-detection} + T\text{-foundation}) \times \text{the number of bounces}$$

On the other hand, the detection time in this case is the expected time to identify an open band, in this case it is irrelevant contrasted to the foundation time, and additionally it is even incorporated into the time of foundation used in few calculations that considers a corner to corner handover or else the vertical handover. Within the writing, it is discovered that the established time is important in misusing of the free frequency band of an additional innovation which is 5 s approximately.

$$T.\text{sensing is less than the } T.\text{establishment, which results to}$$
$$= T.\text{foundation} \times \text{number of jumps.}$$

Crack time: It is the time required of going back for the underlying band frequency. In this way, it is the entirety of the intrusion time and alongside the utilization of a band, realizing that the utilization time contrasts from one band to the other as indicated using the primary client.

2.6   Acquired Outcomes

This segment shows the PU execution influenced by overlaid crafty correspondence of psychological cognitive client as assessed. With respect to the stipulations exhibited in Table 2.2, an ordinary limited mapping and sequence prefix alongside an intonation method provided by the QPSK were likewise prioritized. Additionally, pointing out the simulation intentions about 128 subcarriers accessible to the transmissions was set, while 32 adjoining subcarriers were designated for every 4 dynamic UEs. Featuring this within genuine setting a small stated subcarriers will be accessible to the broadcast since majority are set to convey control signals, whereas others (on edges) are not utilized with the end goal of providing the expected detachment of different groups. However, based on our scenario there is no need of control signals due to the fact that we presume ideal channel acquaintance, though anxiety is created in different frameworks within the neighboring activity groups which are dismissed given that the possibility of this proposition is not even to aggravate in a critical way the framework under investigation.

The examination displayed here was drafted by thinking about dynamic and static distributions of radio assets. In such manner, and going for attaining an execution of

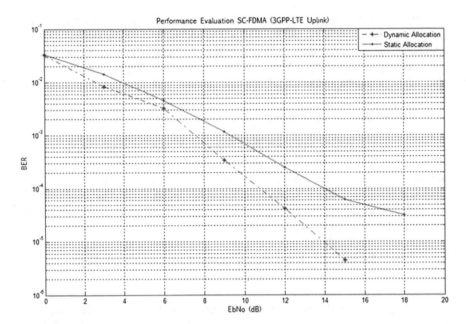

Fig. 2.4 Performance evaluation SC-FDMA

reference, the un-interfered BER essential client was calculated similarly as illustrated in Fig. 2.4.

Each section on every bend shown comprises a consequence for transmitting about almost 14,000 SC-FDMA images subsequent to thinking about a few the E_bNo, which ranges between 0 and 21 dB. Thus the primary factor viewed is a reasonable advantage in including dynamic designation of assets within the framework of the LTE (channel-subordinate scheduler), on the grounds that the BER curve diminishes quicker regarding to the one structured from a static allotment of data transmission. This implies that considering a dynamic allotment of assets the framework quits having blunders with E_bNo that is higher 15 dB, although in other scenarios this happens subsequent to having an E_bNo which is more prominent than the 18 dB.

Along these lines, considering the curves as a reference section, direct data for specific UE in LTE was followed going for inciting a shrewd correspondence occurring within the outrageous circumstances during the time P equals to PSU. Therefore various obstructions were instigated in 2 different techniques; initially it was through overlaying optional information using the subcarriers influenced by invalidity, while the second one was by performing the contrary concept (recognizing the channel tops).

Concerning incited impedance, two limits were chosen purposed to empowering the optional transmission (the edges were picked as the maximum and minimum extracted within the channel by taking into account 14,000 transmissions). Thus the aggregate number of chances distinguished for a prospective situation is exhibited in Table 2.3 below.

Table 2.3 The overall number of concealed opportunities alongside the progress of the UE based on the secondary clients

E_bN_o	Interference BER	Null BER (total)	Peaks BER (total)
0	3.344	4.38	3.909
3	1.441	1.268	1.148
6	4.592	5.848	4.019
9	1.182	2.013	1.485
12	2.481	8.164	2.254
15	6.132	2.891	3.154
18	3.125	5.465	4.124
21	0.1	4.381	1.621

Based on the cognitive radio setting, it is stipulated that this kind of innovation is permitted to some extent bother the essential correspondence. In such manner, it is seen that the reality of overlaying auxiliary information on the most exceedingly dreadful primary subcarriers (as far as channel conditions) leads to a high obstruction UE rate, circumstances which is increasingly obvious as the E_bNo augments. Again, it tends to be seen that executing a resultant transmission covered on the finest subcarriers (channel crests) cannot deform the essential communication past the recently seen magnitude rate, thus when no mistakes are distinguished within the conservative transmissions, therefore the BER of the meddled one shifts upwards to having around one mistake for each every 1,000,000 of bits acquired ($\approx 1 \times 10^{-6}$).

On the other hand, a method for examining the crafty access of the subjective radio transmission via SIR plotting is evidenced in figure eight below relating to transmissions above the crests as the 21 dB is equivalent to E_bNo. This is selected due to the fact that from the examination made previously, the reality of transmitting the best channel conditions is by all accounts the best technique to be trailed by the SU systems. Therefore, when the proposition of co-transmitting within the outrageous channeling conditions was tried from a settled portion of radio assets, a similar sort of investigation was done however this time by considering a channel-subordinate scheduler (dynamic designation of assets). With respect to resultant broadcasts, the threshold limits will be enthused due to prior inclusion of channeling schedulers which is difficult to be distinguished whereas the pinnacles are found easily (nulls from $|H_{PU}|^2 <= 4 \times 10^{-8}$ to $|H_{PU}|^2 <= 8 \times 10^{-8}$, peaks from $|H_{PU}|^2 => 3 \times 10^{-4}$ to $|H_{PU}|^2 => 5 \times 10^{-4}$).

2.6.1   Outcome Analysis

The illustrated chart above explains about the downtime relying upon the quantity of hops executed. Hence it is noted that whatsoever the quantity of jumps similar to the CR is vastly improved compared to the lack of the CR. As indicated by the principal chart, obviously there was loss of 5 s which was required to interface with the

Table 2.4 The overall number of concealed opportunities alongside the progress of the UE based on the secondary clients (Dynamic Allocation)

E_bN_0 (dB)	Interference BER	Null A BER (total)	Null B BER (total)	Peaks BER (total)
0	3.344	3.382	2.382	3.909
3	8.231	9.268	1.268	9.148
6	3.124	2.848	8.848	3.019
9	3.124	1.013	1.013	3.485
12	4.251	8.164	2.164	2.254
15	4.456	3.891	3.891	4.154
18	0	0	0	0
21	0.1	2.381	0	1.002

original band; however, as viewed the CR secured almost 295 s, the time devoid of each interference. In the second chart, we 3 hops were performed, implying the downtime is 5 s × 3 jumps.

The anticipated representation plainly expresses that just the subcarriers experiencing outrageous channel situation are which will be utilized during secondary intentions. Subsequently, the zero as shown in Table 2.3 above provides vivid evidence that no secondary entrance was in actuality and the extraordinary surroundings amidst transmissions cannot satisfy the predetermined limits that were recently laid out [9].

As shown in Table 2.4, various transmissions over the channel nulls were acted upon two times. In the third segment in Table 2.4 (nulls A) affirms that the co-transmission over the profound fading expands the bit mistake tempo within the crucial framework, although the fourth segment (nulls B) shows that when no odds (or an immaterial number) discovered, BER pursues the execution of the un-meddled correspondence. Then again, the last segment demonstrates co-transmission over the station crests prompting the disclosure of more chances, even as in the meantime the execution of the primary user keeps less contorted (the mistakes don't grow up in excess of one request of greatness, and when no open doors are recognized the bit blunder rate remains the equivalent as the regular one).

An intriguing issue that was affirmed using the second analysis is that when the quantity of chances for co-transmitting over the nulls came about to be significantly lesser than the quantity of chances used for co-transmitting over the channel tops, the curves were extensively higher in the primary case. A clarification to this conduct relates to the equalizer installed within the reaction chain that will in general enhance the initiated obstruction situated over the profound fading.

2.7 Summary and Future Work

There is indeed fundamental application and challenge posed on technology due to the high demand of cognitive radio networks. To critically mitigate the issues, CR networks have drastically developed as key initiative that facilitates an innovative access to radio spectrum. This article has presented a unique approach regarding the application of cognitive radio to enhancing the wireless communication connecting the CRMT via the enhancement of QoS for video conferencing. The core contribution of this article is retrieved in machine learning whereby a throughput framework and parameter is critical to performing the categorization of terminals utilized in attaining future events experiences. This implies that it will be possible to identify the time and location to activate CR. The fundamental point-of-focus of CR is its hypothesis, proved on a particular estimated timeframe that connects unique frequency bands and the amount of bands which terminals utilize to control any spoilt connections.

Various approaches applying CR and MAS are demonstrated in the chapter. There are those demonstrating the cooperation between the primary and secondary users, while others offer the cooperation between PUs and SUs and the inclusion of a broker agent who negotiates the spectrum. There is need for future research on the enhancement of wireless links and their reliability to improving the effective service quality of cognitive radio mobile terminal through the integration of MAS. Moreover, it is necessary to research on how the influence of mobility of CR communication will be reduced to develop different predictive mobility models.

References

1. Khan, S., Mitschele-Thiel, A.: Hypernetworks based radio spectrum profiling in cognitive radio networks. EAI Endorsed Trans. Cognit. Commun. 1(2), e5 (2015)
2. Kuiper, D., Wenkstern, R.: Agent vision in multi-agent based simulation systems. Auton. Agent. Multi Agent Syst. 29(2), 161–191 (2014)
3. Tsuji, H., Tsukamoto, K., Suzuki, K., Nagayama, H.: Development of high-speed mobile radio communication systems using 40 GHz frequency band. Radio Sci. 51(7), 1220–1233 (2016)
4. Liu, T., Shao, S., Ye, D., Tang, Y., Zhou, Y.: Visual cognitive radio. Concurr. Comput. Pract Exp. 24(11), 1252–1260 (2011)
5. Wu, Y.: Localization algorithm of energy efficient radio spectrum sensing in cognitive internet of things radio networks. Cogn. Syst. Res. 52, 21–26 (2018)
6. Anandakumar, H., Umamaheswari, K.: An efficient optimized handover in cognitive radio networks using cooperative spectrum sensing. Intell. Autom. Soft Comput. 1–8 (2017)
7. Anandakumar, H., Arulmurugan, R., Onn, C. C.: Computational intelligence and sustainable systems. In: EAI/Springer Innovations in Communication and Computing (2019)
8. Su, H., Moh, S.: A directional cognitive-radio-aware MAC protocol for cognitive radio sensor networks. Int. J. Smart Home. 9(4), 239–250 (2015)
9. Haldorai, A., Ramu, A.: Cognitive social mining applications in data analytics and forensics. Adv. Soc. Netw. Online Commun. (2019)

10. Haldorai, A., Ramu, A., Chow, C.-O.: Editorial: Big Data innovation for sustainable cognitive computing. Mobile Netw. Appl. (2019)
11. Vizziello, A., Amadeo, R., Favalli, L.: Social cognitive cooperation for device to device communications. EAI Endorsed Trans. Cognit. Commun. **3**(11), 152557 (2017)
12. Szydelko, M., Dryjanski, M.: 3GPP spectrum access evolution towards 5G. EAI Endorsed Trans. Cognit. Commun. **3**(10), 152184 (2017)
13. Grace, M., Zhang, H., Nekovee, M.: Editorial: cognitive communications. IET Commun. **6**(8), 783 (2012)
14. Gurugopinath, S., Muralishankar, R., Shankar, H.: Spectrum sensing for cognitive radios through differential entropy. EAI Endorsed Trans. Cognit. Commun. **2**(6), 151147 (2016)
15. Anandakumar, H., Umamaheswari, K.: Energy efficient network selection using 802.16g based GSM technology. J. Comput. Sci. **10**(5), 745–754 (2014)
16. Borra, D., Iori, M., Borean, C., Fagnani, F.: A reputation-based distributed district scheduling algorithm for smart grids. EAI Endorsed Trans. Cognit. Commun. **1**(2), e3 (2015)
17. Guo, W., Huang, X.: Multicast communications in cognitive radio networks using directional antennas. Wirel. Commun. Mob. Comput. (2012)
18. Suganya, M., Anandakumar, H.: Handover based spectrum allocation in cognitive radio networks. In: 2013 International Conference on Green Computing, Communication and Conservation of Energy (ICGCE), Chennai, pp. 215–219 (2013)
19. Anandakumar, H., Umamaheswari, K.: Cooperative spectrum handovers in cognitive radio networks. In: EAI/Springer Innovations in Communication and Computing, pp. 47–63 (2018)
20. Anandakumar, H., Umamaheswari, K.: A bio-inspired swarm intelligence technique for social aware cognitive radio handovers. Comput. Electr. Eng. **71**, 925–937 (2018)
21. Haldorai, A., Ramu, A., Murugan, S.: Social aware cognitive radio networks. In: Social Network Analytics for Contemporary Business Organizations, pp. 188–202 (2018)

Chapter 3
Energy Efficient Network Selection for Cognitive Spectrum Handovers

3.1 Introduction

The wireless communication field has been one of the rapidly growing sectors of the communication industry, as there has been a steady hike in the use of wireless applications. Due to this fact, diverse wireless networking paradigms that are operating under unlicensed spectrum bands have progressively led to congestion of the spectral bands, resulting in spectrum scarcity [1]. However, various investigations on the spectrum have been made by numerous standard telecommunication agencies and it has reported that the major problem is the inefficiency in spectrum usage and scarcity. Thus, a novel and dynamic approach called Dynamic Spectrum Access (DSA) is proposed for spectrum management to address spectrum inefficiencies and usage. This approach should provide wireless access to SUs, allowing them to gain access to unoccupied licensed spectrums [2]. In order to simultaneously guarantee the rights of PUs, an SU can access a spectrum band only if not in use by a primary user.

Since CR involves various functionalities to deliver a Quality of Service (QoS) to its users, energy is being expended in order to perform the required task. As cognitive radio networks possess new technologies when compared to conventional wireless networks and algorithms, additional energy consumption arises [3]. During spectral band, sensing decision-making is required to avoid conflict over the interference between a primary user and Cognitive Radio (CR). The sensed spectrum information should be sufficient for cognitive radio to reach accurate spectral availability. The spectrum sensing technique mainly focuses on energy consumption processes on the CR network to make communications more affordable.

© Springer Nature Switzerland AG 2019
A. Haldorai, U. Kandaswamy, *Intelligent Spectrum Handovers in Cognitive Radio Networks*, EAI/Springer Innovations in Communication and Computing,
https://doi.org/10.1007/978-3-030-15416-5_3

3.2 Literature Survey

The [4] considered Base Station Controller (BSC) as an interference management technique for WiMAX. A cooperative transmission scheduling scheme with minimal complexity, user feedback, and information exchange between the base stations was proposed by the authors. This work also evaluated the proposed technique with non-cooperative schemes, similar in complexity via Monte Carlo simulations. BSC also provides a smart solution for mitigating the CCI and increasing spectral efficiency of the system, based on frequency reuse.

The [5] proposed energy efficiency problems like (1) reliability of sensing, (2) throughput, and (3) delay of SU. Reliability of sensing can be measured by the probability of detection and probability of false alarm. In probability of detection, it is sensed as busy when the channel is busy. The highest probability of detection is that SU can catch a PU communication more accurately. In probability of false alarm, it is sensed as busy when the channel is idle. Hence, the throughput amount of data should be satisfied and the delay of SU should not be too lengthy.

The [6] proposed the benefits of channel bonding in 802.11n. The protocol was influenced by network factors such as interference and loss. Therefore, it is clear that channel management solutions in 802.11n must understand the behavior of channel bonding in order to make intelligent decisions to assign bandwidth in the network.

The [7] proposed a cluster and forward based distributed spectrum sensing scheme for energy utilization where cluster heads were formed during processing of spectrum sensing from new secondary users. This technique was only able to deal with energy efficiency.

The [8] studied the problem of joint power control and beam forming, to maximize the network utility function. They proposed a convergent alternating optimization algorithm that was applicable to decentralized wireless networks. The paper studied various theories about the stochastic approximation and algorithm convergence properties.

The [9] studied a set of detected spectrum bands that can be temporarily used by each node in a dynamic spectrum access network, to form a topology by selecting spectrum bands for each radio interface of each node. A layered graph to model the temporarily available spectrum bands was proposed in the paper, and the model was used to develop effective routing and interface assignment algorithms to form near optimal topologies for Dynamic Network Access (DSA). This model provided solutions for DSA networks with static link properties. A fixed channel approach was considered, and the radio interfaces were assumed to tune in a wide range, but operate on a limited and smaller range at a specific time. Specifically, a radio interface was assumed to operate on only one channel at a specific time.

CR [10] considered the existence of multiple PUs performing cooperation with multiple SUs in the network. Specifically, the transmission of PUs was divided into different frames, and different pairs of PU and SU perform cooperation over different frames to maximize the network utility. Cooperation for multi-channel CR networks was investigated, where multiple PUs operating over different channels cooperate with different SUs, simultaneously to maximize the network utility. Maximum weight matching was utilized to coordinate the support between multiple PUs and multiple SUs.

The average energy cost of the SU [11] includes spectrum sensing, channel switching, and data transmission. This was optimality achieved by considering two fundamental tradeoffs, such as sensing/transmission and wait/switch tradeoff. An efficient, convex optimization procedure was developed to solve the optimal values of the sensing slot duration and channel switching probability.

The [12] consigned over the interference management issue that occurs in air interfaces between low and high data rate personal area network on the same multimodal device. IEEE 802.15.4 and ECMA-368 standards are considered in this work, where a collaborative coexistence is proposed among complementary wireless personal area networks. The work also evidences that the alternating wireless activity mechanism efficiently synchronized the home register and location register air interfaces.

The [13] had taken into account that active and inactive PUs coexist on the primary network. The paper proposed two simple cooperation frameworks, capable of stimulating both active and inactive PUs, and SUs.

The active PUs and SUs can relay PUs' packets and obtain transmission opportunities as a reward. For inactive PUs, neighboring SUs can lease spectral bands from inactive PUs and perform cooperative communications with each other.

The literature survey has analyzed and given a perspective that research works have previously only focused on optimizing energy efficiency over CR networks. Whereas, certain attention could be given towards its solutions for enhancing energy efficiency, hence, our proposed research is focused on improving and addressing energy efficiency issues in CR networks.

3.3   Energy Efficient Network Selection

Energy efficiency in CR networks has become a concern to many stakeholders, in the rapid development of wireless communications. As CR networks consist of energy demanding components like base stations and various terminal nodes. The lifetime of the network is fully dependent on the energy consumed by the above components in the various stages of communication [14]. This paved the way towards network operation and design aspects of CR regarding energy efficiency.

Energy utilization techniques in CR networks are as follows:

- Perform energy saving at various levels of CR activities.
- Reduce the interference to attain high signal noise ratio (SNR) with the same transmission power.
- Increase the speed of sensing in order to save energy in periodic sensing.
- Work under 802.16g, protocol based active/inactive modes.

3.3.1 Significance of Energy Consumption in CRN

As the rate of wireless devices increases drastically, demand for energy efficient devices constantly increases. Nowadays, various government agencies, service providers, network device manufacturers and users concern over the energy redemption issues of wireless devices [15]. The significance of optimizing energy is concerned with design, green communications policy, monetary cost, and end user's gratification and fulfillment for CR networks.

More heat is expelled when the CR user is in communication, leading to the malfunctioning of devices. In order to reduce temperature, a large cooling system is required, but this is not easily applicable in mobile devices; but design issues must be considered for wireless communication systems to have high energy efficiency.

Environmental scenarios like the greenhouse effect are a source of concern to various government organizations and propose standards to necessitate wireless devices to be more energy efficient. Therefore, energy consumption is managed by devising protocols towards the optimization of energy in CR networks.

Nowadays, radio base stations consume around 82% of energy in mobile telecommunication systems [16]. A huge amount of energy is required by each base station to transmit and receive wireless data, and the use of the energy efficient protocol reduces heat from these wireless components.

Wireless devices are used as network terminals during mobility in CR networks, which are lightweight and have a long battery life. New research on battery power technology can extend the battery capacity and simulate the energy efficient protocol, which will be used to make cognitive radio devices run efficiently under minimal power.

3.3.2 Energy Efficient Sensing

In wireless network environments, single base station energy consumption rate varies between 0.5 kW up to 2.0 kW, so base stations require more energy efficient CR networks [17].

There are many factors that enhance the energy utilization in CR networks, with most of them created due to spectrum sensing. An efficient cognitive transceiver is necessary to attain a wide range spectrum, at the same time detecting primary signals that are diverse and frequency dependent. This demands antenna, power amplifiers, and analog to digital conversion units of RF front-end, to use more energy.

A certain amount of energy is needed by a CR user in order to deliver secondary data via available spectrum holes. This demands high processing power by each signal processing unit, to effectively analyze the pre-sensed spectrum to make a decision with minimal delay over total energy consumed at the network [18]. As a result, the energy utilized by a high-power CR amplifier is nearly 70% of the total energy consumption.

If the power consumption at CR terminal can be managed, then the energy optimization in the CR networks will be increased. CR users consume a certain amount of energy during different stages of the user's activity like: (1) transmission state, (2) collision state, (3) idle state, (4) sleep state, (5) channel scanning, and (6) back-off states.

3.3.3 Maximize the Energy Efficiency

In order to attain high energy efficiency, a significant amount of hurdles must be overcome in CR networks. Each stack of the CR network consists of several sets of components, diverse protocols, and varied mechanisms. Many approaches are widely concerned with power consumption in CR during different modes of activities, and some of them are working on reducing the rate of interference, thereby minimizing the error rate during retransmission [19].

Other measures for achieving high energy efficiency in CR networks are concerned towards the improvement of the transmission rate that leads to reduced total transmission time. The process of green energy collection is also a means of enhancing energy, since it can serve as a source of power supply, restricting CR to depend on other power sources.

3.3.4 Intelligent Channel Selection

The channel management focuses on the usage of a dedicated, interference free channel. Channel management can be broken down into two types: (1) central control and (2) distributed control. The central control mechanism is aimed at broadcasting control channel messages to the entire network, embedded with information on the suitable channel to be used [20]. Distributed control on the other hand does not depend on the node's coordination; however, all devices must operate with an identical approach.

3.3.5 Intelligent Protocol Selection

The efficiency of various IEEE 802.x protocols has been analyzed and through a detailed survey which is stated in Table 3.1, it is proposed to utilize and adapt IEEE 802.16g in order to support low battery outage and also increased data transmission under minimal computational complexity.

The proposed protocol IEEE 802.16g is used for direct transmission technique by allowing enough provisions for the best effort QoS data streams which do not require any higher level of service.

Table 3.1 Comparative analysis of various energy efficiency techniques

Author	Technique used	Disadvantage
Xin (2005) [28]	Dynamic network access (DSA)	Utilizes only fixed channel approach for radio interference
Cavalcanti et al. (2007) [29]	Distributed coordinated function protocol 802.11e	Ensures power reduction only for direct link setup
Ying-Chang Liang et al. (2008) [30]	Probability of detection	It is difficult to predict false alarm when delay of secondary users is too long
Chandra et al. (2008) [31]	Collaborative spectrum sensing protocol	Detect vacant spectral bands only when channel width and data rate is high
Shrivastava et al. (2008) [32]	IEEE 802.11n protocol	Intelligent decisions not feasible
Monti et al. (2010) [33]	IEEE 802.15.4 and ECMA-368	Collaborative coexistence proposed but only for personal area network
Song and Zhang (2010) [34]	Distributed spectrum sensing scheme	Focus only on energy efficiency but not handover
Tokel and Aktas (2010) [35]	Base Station Controller (BSC)	Requires user feedback
Feistel et al. (2010) [36]	Joint power control and beam forming	Not focused on centralized wireless networks
Liu et al. (2013) [37]	Cooperative communication between PU and SU	Requires active and inactive PUs to coexist, but not focus on dynamic PU and SUs

3.4   Proposed Energy Efficient Network Architecture

3.4.1   Cooperative Active/Inactive Mode

Switching to the base station sleep state is a means of enhancing the energy efficiency of CR network. Though the sleep mode can cause significant reduction in delayed service and minimal QoS for CR users, energy consumption can be effectively minimized with proper protocol [21]. During Base Station (BS) sleep state, its transmissions are turned off to reduce traffic load and also the power amplifier inside the base station radio blacks out.

Figure 3.1 depicts the initiation of energy efficient mode when users are not able to identify signal or base station for transmission in order to perform handover.

The processes are presented below:

- Scan: Perform search for nearby base stations to activate service activation.
- Efficient network selection: When it identifies a suitable BS, it activates.
- Idle state: When there is no accessible network, it goes into an idle state. This state is also called "sleep mode." It remains in this state until it identifies a BS and enters an efficient network selection mode.
- Select BS/mobile node: The channel is allocated to available secondary network for required mobile terminal in this mode.

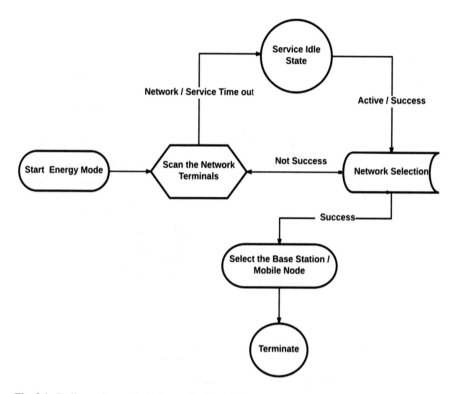

Fig. 3.1 Radio on time optimization with sleep state

Figure 3.2 depicts the same process in Fig. 3.1, but varies in the idle state, by indicating a lookup state and finally enters selected BS state.

Lookup state: Activation of base station sleep mode is through one of two ways: deep sleep phase and micro sleep phase.

- In micro sleep state, the transmission of radio base station is seized to a minimum duration and is needed to wake up immediately.
- In the deep sleep state, the transmission of radio base station is shut down for an extensive time and some transmit states are completely terminated.

3.4.2   IEEE 802.16g Network Selection Model

In the method of cooperative spectrum sensing, types of CR networks can be categorized into centralized and decentralized networks [22].

- In the centralized cooperative spectrum sensing, the network coordinator collects local observations from the nodes and then decides on the spectrum occupancy, using decision fusion rules and informs the nodes on which channel to access.

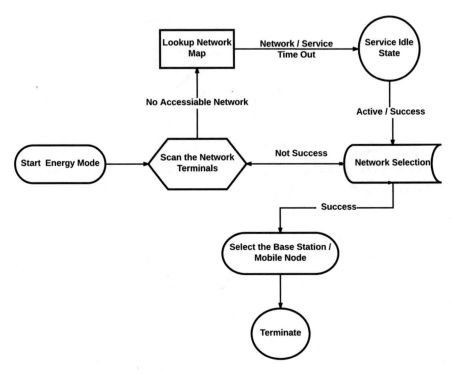

Fig. 3.2 Radio on time optimization without sleep state

- In decentralized cooperative spectrum sensing, nodes within a CR network exchange their local spectrum sensing results among themselves without requiring a backbone or network coordinator.

Figure 3.3 shows the architectural model of IEEE 802.16g protocol where the wireless communication network works with a physical and MAC layer. This design is required when activating an efficient network selection.

The management Service Access Point (SAP) has the following features to increase energy efficiency during network selection:

- SAP controls PHY, DLL, and network layers, thereby supporting "Efficient Data Transmission" during the inactive mode.
- A centralized control is provided in management SAP, which uses "fusion rules" to increase SNR and reduce power attenuation.
- To improve handover by satisfying (1) low battery outage and (2) high data transmission.

System performance is monitored in terms of system throughput, as the packet generation rate is being varied for the traditional IEEE 802 protocol standards during spectrum sensing and allocations.

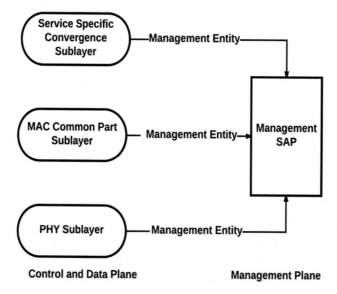

Fig. 3.3 802.16g Architectural model

3.5 Energy Efficient Spectrum Sensing

3.5.1 Energy Efficient Multi-antenna System

The antenna systems influence the efficiency of energy usage in a CR network. A Direct Transmission (DT) is an effective as well as user-friendly transmission, but it has only one user at each end of the wireless link [23]. This is a low-level transmission when compared to the Cooperative Transmission (CT), which helps in expanding the capacity of the wireless network with additional antennas within the networks. With the aid of fast transmission process, the CT antenna method has the capacity to increase the efficiency of the energy usage in CR network.

The time and energy consumed by radio and transmitter is less during the faster transmission process. The spectrum space will be made available for the remaining traffic when the sensing process of the spectrum is faster. Energy efficiency in CT is improved by deactivating a few antennas, in certain fast transmission situations to reduce the required signals in the channels.

CR networks have a CT antenna system, along with adaptive transmission techniques with prior knowledge about channel transmission, to enhance the energy and spatial efficiency, because precise information on the wireless channel environment to the transmitter and the receiver is mandatory [24]. This data is sent in the mode of reference symbols, which are accepted by the receiver. This ensures that the CT mode has the advantage to coordinate more users, for the same amount of available resources.

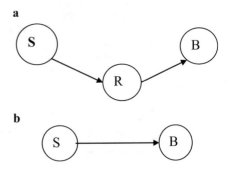

Figure 3.4a is a representation of direct transmission of data to the base station
from the Secondary Mobile Terminal (SMT), whereas Fig. 3.4b represents the
cooperative transmission from SMT to the base station with the help of Home
Mobile Terminal (HMT). In an ideal state, the secure information of the HMT,
which includes the channels, its energy levels and the conditions, is mutually shared
to both HMT and the SMT. In a real-time scenario, the secure information is not
shared between HMT and SMT.

A Mobile Terminal (MT) does not possess any information about the remaining
MT's channel condition and battery levels [25]. Hereby, it is assumed that the entire
channel gain and the HMT's battery level are independent, and the measured
reservation utility indicates the minimum benefit level of cooperation. The expected
energy level of SMT is considerably higher, to attain higher reservation utility for the
HMTs. The HMT channels are identified based on this process.

The channel allocation to SMT by HMT and the access gained from the back of
channels by the SMT is done with the consideration of the set of idle MT which is
used to point out within the short-range communication at distance "d" from the MT,
as total number of MT in the HMT channel is H_i where $|H_i| = N_i$. In this case N_i
follows a Poisson distribution and its Probability Density Function (PDF) value is
measured. If $N_i \geq 1$, then SMT operates in cooperative transmission mode by
selecting an HMT for data transmission. The channel gain between SMT and HMT
is calculated to select the HMT. The difference between the cost of energy and utility
function calculated based on the HMT is determined [26]. A time slot is required to
prepare access the channels by the SMT. These time slots have symbols for the data
that is to be transmitted and these data symbols are transmitted in every time slot. The
quantities of symbols that are permitted through the transmission per time slot are
brought down to normal transmission without any loss of information or data.

3.5.2   Energy Efficient Interference Management

The level of noise that increases in channel denotes the interference and leads to
reduced SNR to the transmission channel. When signal transmission need to be

completed over high interference channels, the power consumption will be higher as they have to rise over the interference, in turn making the signal interrupt other signals. In these cases, CR is expected to solve the issue by avoiding those channels which might cause interference [27]. Though the CR would not be expected to resolve the interference problems arising from these signals, other devices may be affected by the signals produced from these devices. At this juncture, the IEEE 802.16g protocol will efficiently manage any interference from the device in use and avoid interfering with other devices.

3.5.3   Cooperative Sensing and Optimization

Within the spectrum hole, there is one BS serving "K" MTs depicted by the set $K = \{1, 2, \ldots k\}$. Taking into consideration of all the MTs uplink data transmission, an assumption of the narrow-band block fading channel model is made. The channel baseband coefficient obtained from MT $K \in \kappa$ to the BS is denoted as h_k, and it is followed by a simplified channel model which incorporates the large-scale power attenuation to that of the loss exponent $\alpha > 2$ and the small-scale Rayleigh fading. Then, the channel coefficient h_k is expressed in Eq. (3.1)

$$h_k = \begin{cases} \bar{h}_k \sqrt{G_0} \left(\frac{r_k}{r_0}\right)^{-\alpha}, & r_k > r_0 \\ \bar{h}_k \sqrt{G_0}, & \text{Otherwise} \end{cases} \tag{3.1}$$

where

- r_k is the distance between MT $K \in k$ and the BS.
- r_0 is a reference distance.
- $\bar{h}_k \sim G(0, 1)$ is a random variable (RV) with zero mean and unit variance modeling.
- G_0 denotes the path loss constant between MT and BS at distance r_0.

The channel power gain between MT k and BS is shown in Eq. (3.2)

$$g_k = |h_k|^2 = n_k G_k, \quad k \in K \tag{3.2}$$

The data traffic or transmission is initiated independently by the MTs. In a simplified channel model, which incorporates a large-scale power, attenuation will be described in Eq. (3.3)

$$G_k = \begin{cases} G_0 \left(\frac{r_k}{r_0}\right)^{-\alpha} & r_k > r_0, \quad k \in K \\ G_0, & \text{otherwise} \end{cases} \tag{3.3}$$

Data rate (D_k) is made normal within the accessible bandwidth of the MT for the transmission of every symbol with E_k energy and the obtainable data rate for MT is identified in bits/s/Hz (bps/Hz) as in Eq. (3.4)

$$D_k = \log_2\left(1 + \frac{g_k E_k}{\sigma^2}\right) \tag{3.4}$$

where σ^2 is denoted as the power of noise at BS receiver.

The energy per symbol that is used to transmit with a data rate D_k is shown in Eq. (3.5)

$$E_k^{(D,S)} = \frac{\sigma^2}{g_k}\left(2^{D_k} - 1\right) \tag{3.5}$$

3.5.4   Direct Transmission (DT Mode)

The MTs transmit their data directly to the BS in transmission mode with a normal data rate D_i and compute the energy level (E_i) per symbol to send data from MT to BS at a data rate D_i as denoted in Eq. (3.6)

$$E_i^{(D,S)} = \frac{\sigma^2}{g_i}\left(2^{D_i} - 1\right), \quad i \in K_s \tag{3.6}$$

3.5.5   Cooperative Transmission (CT Mode)

During the CR cooperative transmission mode, the SMTs are coordinated in helping the mobile terminal within the short-range distance d, and the HMT will be assisting in relaying the data of SMTs to the BS. The set of Ideal Mobile Terminal (IMT) and the general MTs are considered as HMT "H_i" at distance "d" from the SMT derived in Eq. (3.7)

$$\mu N_i = (1 - \rho)\lambda \pi d^2 \tag{3.7}$$

Here λ denotes the spatial density and 1-ρ denotes the MTs probability.

The probability mass function of N_i is directly proportional to distance d and is stated in Eq. (3.8)

$$P_r(N_i = n) = \frac{\mu^n N_i}{n!} e^{-\mu N_i}, \quad n = 0,1, \dots, i \in K_S \tag{3.8}$$

where

μN_i denotes the average number of HMT for SMT.

If $N_i = H_i = 0$, then SMT will enter the direct transmission mode and transfer data directly to the BS. But when $N_i, H_i > 1$, then SMT terminal will try to find a HMT and transmit the data through HMT.

The CT data rate must be split in Eq. (3.9)

$$D_i = D_I^{(S)} + D_i^{(R)} \tag{3.9}$$

The energy is explored in case SMT fails to find a better HMT to relay its data at a specific condition. Here $D_I^{(R)}$ denotes data rate of cooperative transmission SMT and HMT. During CT mode, SMT disseminates its data to the selected HMT and it decrypts the received data and forwards it to BS. The amount of power gain among SMT and HMT is $g_i = n_j G_i$, where n_j is the short-term Rayleigh fading of the channel power and it is independently distributed among MTs, and the energy consumption during data transmission is given in Eq. (3.10)

$$E_j^{(C,R)} = \frac{\sigma^2}{g_j} \left(2^{D_i^{(R)}} \right), \quad j \in H_i, \quad i \in K_s \tag{3.10}$$

3.5.6   Cooperative Transmission Protocol

A cooperative network should maintain its communication protocol for the purpose of data transmission as follows:

- If any SMT has data that has to be transmitted to the BS, it will identify the data mode as CT or DT as per the desire of the SMT which may be either complete or incomplete data.
- In a DT mode, the SMT will be able to transmit directly all the data to the BS. On the other hand, if the SMT selects the CT mode, the source MT has to make an announcement about the data rate procedures to all of the HMTs connected to the BS.
- If the request is accepted by any of the MT, an acknowledgement will be sent to the SMT and the data will be relayed to the BS.
- If more than one MTs acceptance is found, the SMT will have to decide upon the best MT suitable for the condition. Later the chosen HMT will aid in relaying data to its BS.

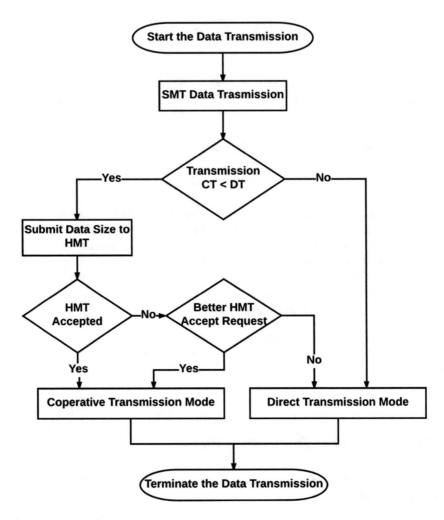

Fig. 3.5 Overview of cooperative communication protocol

Figure 3.5 reveals the flow of the CT protocol and if more than one MT is willing to do the job and accepts the request, then the SMT will have to filter and choose the best by means of CT protocol.

3.5.7   Cooperative Sensing and Data Fusion

There are "K" CRs nearby which are able to join a cooperative sensing network and each CR sends one bit decision to $K_i(I = 1, 2,\ldots, K; K_i = 0$ or $1)$ Data Fusion Center (DFC). The DFC receives K binary decisions and makes a final decision using the binary data. If the summation of the received bits is less than decision threshold λ, DFC decides that PU is not transmitted. If the summation of receiving bits is greater than or equal to decision threshold λ, then DFC decides PU is transmitted as shown in Eq. (3.11)

$$D_{\text{FC}} = \begin{cases} \sum_{i=1}^{k} K_i < n, & H_0 \\ \sum_{i=1}^{k} K_i \geq n, & H_1 \end{cases} \tag{3.11}$$

where

H_0—Absence of primary user
H_1—Presence of primary user
Signal of time "t" and bandwidth "w" detected.
H_1 is detected during spectrum "S" and compared with detection threshold "λ" derived in Eqs. (3.12) and (3.13)

$$H_0 : X(t) = n(t) \tag{3.12}$$

$$H_1 : X(t) = h(t) + n(t) \tag{3.13}$$

where

$t = 0, 1, 2 \ldots N$
$X(t) =$ Secondary user's received signal
$H(t) =$ Primary user signal
$n(t) =$ Additive White Gaussian Noise (AWGN)
A comparison is made on diverse DFC by using FCC recommendations and a handover mechanism based on current state and detection threshold is obtained energy prediction which is shown in Table 3.2.

Case 1: The system prediction declares H1 and the PU is actually present (H1) (true prediction) as a result, $\lambda = \lambda \rho$new. This will increase the value of P_d compared to the fixed threshold method.

Case 2: The system prediction declares H1, but actually the PU is absent (H0) (false prediction) as a result, $\lambda = \lambda \rho$new. This will increase the value of Pfa compared to the fixed threshold method.

Table 3.2 DFC under proposed CT protocol

Scenario	Next state	Current state	Detection threshold λ_{new}	Result	Energy prediction
1	H_1	H_1	λ/ρ	P_d increases	Good
2	H_+	H_0	λ/ρ	P_d increases	Poor
3	H_0	H_0	$\rho\lambda$	P_d increases	Good
4	H_0	H_1	$\rho\lambda$	P_d increases	Poor

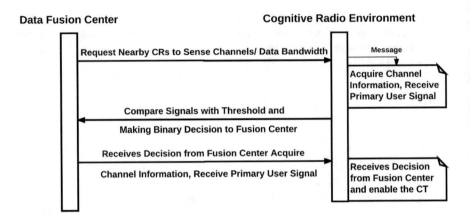

Fig. 3.6 CR cooperative sensing network diagram

Case 3: The system prediction declares H0 and the PU is actually absent (H0) (true prediction) as a result, $\lambda = \rho\lambda$new. This will decrease the value of Pfa compared to the fixed threshold method.

Case 4: The system prediction declares H0 but the PU is actually present (H1) (false prediction) as a result, $\lambda = \rho\lambda$new. This will decrease the value of P_d compared to the fixed threshold method.

Case 1 and 3 are a representation of cases where PU does not have the intention to change the status for an extended period of time and the cases 2 and 4 indicate that the PU changes its status from ON to OFF or from OFF to ON. The proposed scheme outperforms the conventional fixed threshold scheme in the PU steady state situations, wherein the PU does not have any sudden changes in its status.

Figure 3.6 shows the process of CR cooperating sensing network where a request is sent by the DFC to sense if there are any channels available for usage. The CRs obtain the available channel information and the PU signal, thereby, comparing the signal threshold and received for decision-making. Any final decision of the available or free channels of the PU is done through the DFC and later received by the CRs which will finally enable the CT.

3.6 Experimental Results

3.6.1 Energy Simulation Setup

The comparison of single relay selection and multi-relay selection is explored under the proposed technique for direct and cooperative transmission mode with existing models, and the result is simulated under 100×100 m^2 area with data rate $D_i = 16$ bps/Hz. Here the position of the MTs is simulated in the iteration.

- The simulation performed using the proposed techniques was implemented using Matlab 7.10.0 (R2010a) in a 64-bit computer with a core i3 processor (clock speed of 2.8 GHz) and 4 GB RAM.
- The signal bandwidth is configured as 7.56 MHz and locating at the central radio frequency of 720 MHz.

(As per Federal Communications Commission (FCC) standard, the receiver shall be capable of receiving all channels within the frequency range of VHF and UHF. The signal bandwidth of each channel is 8 MHz and its effective bandwidth of the signal is 7.56 MHz. Hence the central frequency of 720 MHz (band pass filter) is used to split the signal into various channels).

- A total under **100 × 100 m²** area in search space was made to run by setting $D_i = 16$ bps/Hz as the data rate.
- Let the number of samples (mobile terminals) be considered as $K = 100$ in the search space.
- The probability of MTs for data transmission assigned as $\rho = 0.2$.
- Number of home mobile terminals fixed as $\mu N_i = 1.2$. (HMT gains the access from the back of channels of SMT).

Figure 3.7 shows the MTs' position which has been functioning off the process raising battery levels of MTs, which are generated on a uniform scale of [0, B_{max}]. The positions of MTs during the start of each time slot are regenerated evenly within the specified area. During this process, there is a possibility of overlapping among the neighboring MTs from different source MTs. There is a chance of one

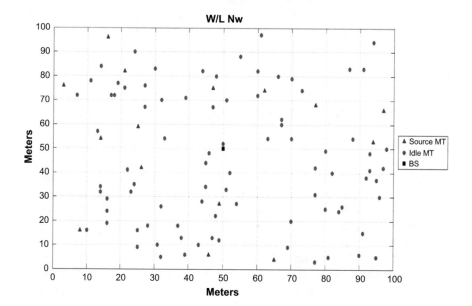

Fig. 3.7 Single relay selection of MT with |K|=100 MTs

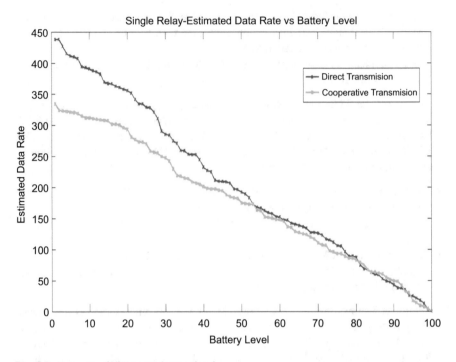

Fig. 3.8 Data rate of MT versus battery level

neighboring MT being linked together with two source MTs. To avoid this condition, there is a regeneration of the source MTs positions required. The batteries of certain MTs are drained out during system process, and if any MT states a battery outage then its operation ceased from data transmission.

3.6.2   Energy Consumption Result Analysis

3.6.2.1   Single Relay Model

The energy efficiency is analyzed for computing battery outage and data transaction. The efficiency of the battery level is compared in direct mode and cooperative mode by simulating results over single relay selection. Figures 3.8 and 3.9 illustrate the comparison of data rate vs. battery level and average battery level vs. time, respectively.

The number of neighboring MTs N_k for source MT $k \in KS$ is generated based on Poisson distribution with $\overline{\mu}N_k = 2$, whereas the rate of source MTs data transmission is $D_k = 4$ bps/Hz. The battery levels of MTs are B_j, j which is uniformly generated over $[0, B_{max}]$ and a short-term Rayleigh fading $\eta_j, 2H_k$ is generated. A minimum of over 1000 independent realizations is averaged as the minimum energy costs in order

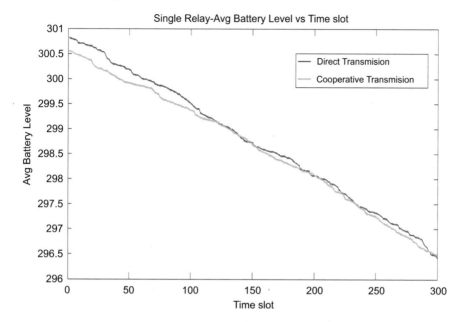

Fig. 3.9 Average battery levels $\sum_k B_k |K|$ of the MTs over time slot

to get a precise cost of energy, which is used, for comparison. The same cost of DT mode transmissions is expected to acquire with the same average of 1000 independent realizations.

Figures 3.8 and 3.9 depict the simulation results of the proposed cooperative relaying scheme which shows superior performance when compared to direct transmission. Whenever the level of the battery from source MT shows a transparent low, cooperative relaying is very effective. This condition prevails because whenever the source MT's battery level is high and the direct transmission cost is low, the tendency to receive from other MTs is rarely noticed.

3.6.2.2 Multi-Relay Model

The multi-relay selection and its simulation are analyzed using (1) energy efficiency of the battery and the data transaction and (2) efficiency of battery level in direct mode and cooperative mode. In a multi-relay selection simulation, probability of overlapping of the HMT towards many SMTs is higher and any one particular HMT will be able to relay between two entirely different SMTs. In order to avoid this awkward situation, the function $\mu N_i = (1 - \rho)\lambda \pi d^2$ is used for regenerating the MTs position and if there exists any overlap between the HMTs, then the average number of HMT is $\mu N_i = 1.2$ and maximum energy for any time slot is 3 J.

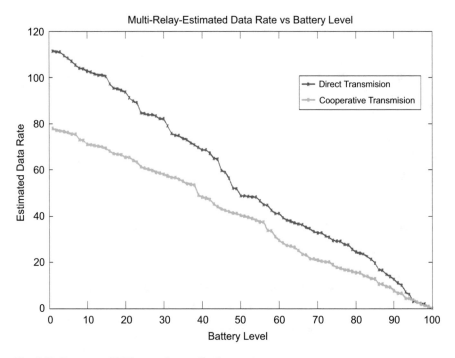

Fig. 3.10 Data rate of MT versus battery level

Figure 3.10 shows the comparison of estimated data rate vs. battery level and Fig. 3.11 shows the comparison of average battery level vs. time slot in both direct and cooperative transmission modes.

The above simulation indicates that the proposed cooperative relaying method functions more efficiently than that of direct transmission. This proposed protocol can be used in an effective manner to enhance the battery levels to an average value of MTs over time for the multi-relay model.

3.7   Discussion

Table 3.3 shows the best cases considered for efficiency of energy during the period of battery outage from 100 random MTs throughout a period of 300 time slots. It is shown that the number of MTs were found to have their batteries drained during DT. In other conditions when cooperation is there, the levels of their batteries seemed to be on the higher end, than that of direct transmission through distribution. Many of the MTs under the wing of cooperative relaying seem to remain under the low battery area of 0–20 J; in reality. their batteries are not completely empty. The MTs are able to get sustenance from other MTs when their own battery levels are low, by which they maintain their battery levels without drying out. On the other

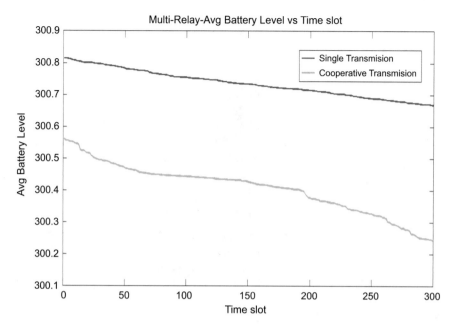

Fig. 3.11 Battery levels $\sum_k B_k |K|$ of the MTs over time slot

Table 3.3 Energy efficiency during 300 time slots

Single relay model		Multi-relay model	
Direct transmission	Cooperative transmission	Direct transmission	Cooperative transmission
1.909915	0.555562	1.980689	0.686610
2.080958	0.698975	2.137103	0.516680
1.853760	0.437461	1.697227	0.730340
1.724548	0.741290	1.960547	0.837098

hand, in the case of direct transmission, many of the MTs of any particular region have their batteries completely drained and empty because of lack of facilitating from nearby relaying MTs.

The proposed protocol of IEEE 802.16g CT has proven to be one of the best protocols, allowing enough provisions for the best effort QoS data streams which do not require any higher level of service. This functions on the basis of space available system. Figure 3.12 describes the comparisons between the proposed technique and the prevailing method. This makes an indication that the CT model is 61.9% better and has higher efficiency techniques to be used in the reduction of battery shortage or complete drying out of the MTs. It also shows an increased efficiency in energy management with a bare minimum level of the battery in the given time slot.

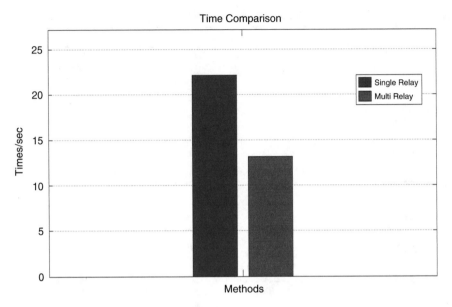

Fig. 3.12 Singe relay vs. Multi-relay model

3.8 Summary

It can be observed that when compared with the standard case of DT, the proposed cooperative relaying performs better in terms of increased communication and reduced battery outage. The reduction of the battery outages from the results displays the effective energy saving design of the proposed protocol for those MTs in low battery level. The proposed CT protocol scheme improves the MTs data transmission during uplink in terms of reliable communication and sustainable battery by satisfying following requirements: (1) Reduced interference with high SNR, (2) Increased speed of sensing and energy consumption with periodic sensing, and (3) Efficient handover at BS inactive (both deep sleep and micro sleep) state.

References

1. Anandakumar, H., Umamaheswari, K.: An efficient optimized handover in cognitive radio networks using cooperative spectrum sensing. Intell. Autom. Soft Comput. 1–8 (2017)
2. Anandakumar, H., Arulmurugan, R., Onn, C.C.: Computational intelligence and sustainable systems. In: EAI/Springer Innovations in Communication and Computing (2019)
3. Celebi, H., Arslan, H.: Utilization of location information in cognitive wireless networks. IEEE Wirel. Commun. **14**(4), 6–13 (2007)
4. Beibei, W., Liu, K.J.R.: Advances in cognitive radio networks: a survey. IEEE J. Select. Top. Signal Process. **5**, 5–23 (2011)

5. Suganya, M., Anandakumar, H.: Handover based spectrum allocation in cognitive radio networks. In: 2013 International Conference on Green Computing, Communication and Conservation of Energy (ICGCE), Chennai, pp. 215–219 (2013)
6. Qing, Z., Sadler, B.M.: A survey of dynamic spectrum access. Signal Process. Mag. IEEE. **24**, 79–89 (2007)
7. Lu, L., Zhou, X., Onunkwo, U., Li, G.Y.: Ten years of research in spectrum sensing and sharing in cognitive radio. EURASIP J. Wirel. Commun. Netw. **2012**, 28 (2012)
8. Barra, C., Zotti, R.: Bank performance, financial stability and market concentration: evidence from cooperative and non-cooperative banks. Ann. Publ. Cooper. Econ. (2018). https://doi.org/10.1111/apce.12217
9. Tie, X., Hong, P., Xue, K., Tang, H.: A handover mechanism based on cooperative diversity. J. Electr. Inform. Technol. **33**(5), 1178–1185 (2011). https://doi.org/10.3724/sp.j.1146.2010.00818
10. Huang, C., Zhang, W., Li, K., Huang, B., Dai, B.: Hierarchical network coding based cooperative handover mechanism in wireless internet of things. J. Electr. Inform. Technol. **35**(1), 147–150 (2014). https://doi.org/10.3724/sp.j.1146.2012.00468
11. Yang, J., Ji, X.: An improved inter-domain handover scheme based on a bidirectional cooperative relay. Cybernet. Inform. Technol. **13**(4), 127–138 (2013). https://doi.org/10.2478/cait-2013-0059
12. Papadaki, K., Friderikos, V.: Optimal vertical handover control policies for cooperative wireless networks. J. Commun. Netw. **8**(4), 442–450 (2006). https://doi.org/10.1109/jcn.2006.6182792
13. Jones, R., Veendorp, E.: Cooperative moves in a non-cooperative game. Glob. Business Econ. Rev. **7**(1), 25 (2005). https://doi.org/10.1504/gber.2005.006917
14. Haldorai, A., Ramu, A., Murugan, S.: Social aware cognitive radio networks. Soc. Netw. Anal. Contemp. Business Organ. 188–202 (2018)
15. Anandakumar, H., Umamaheswari, K.: Supervised machine learning techniques in cognitive radio networks during cooperative spectrum handovers. Clust. Comput. **20**(2), 1505–1515 (2017)
16. Cabric, D., Mishra, S.M., Brodersen, R.W.: Implementation issues in spectrum sensing for cognitive radios. Conference Record of the Thirty-Eighth Asilomar Conference on Signals. Syst. Comput. **1**, 772–776 (2004)
17. Yucek, T., Arslan, H.: A survey of spectrum sensing algorithms for cognitive radio applications. IEEE Commun. Surv. Tutor. **11**(1), 116–130 (2009). https://doi.org/10.1109/SURV.2009.090109
18. Shen, J., Jiang, T., Liu, S., Zhang, Z.: Maximum channel throughput via cooperative spectrum sensing in cognitive radio networks. IEEE Trans. Wirel. Commun. **8**(10), 5166–5175 (2009)
19. Zhao, Z., Peng, Z., Zheng, S., Shang, S.: Cognitive radio spectrum allocation using evolutionary algorithms. IEEE Trans. Wirel. Commun. **8**(9), 4421–4425 (2009)
20. Haldorai, A., Ramu, A.: Cognitive social mining applications in data analytics and forensics. In: Advances in Social Networking and Online Communities (2019)
21. Sadreddini, Z., Güler, E., Çavdar, T.: PSO-optimized instant overbooking framework for cognitive radio networks. In: 2015 38th International Conference on Telecommunications and Signal Processing (TSP), Prague, pp. 49–53 (2015)
22. Wang, G., Guo, C., Feng, S., Feng, C., Wang, S.: A two-stage cooperative spectrum sensing method for energy efficiency improvement in cognitive radio. In: 2013 IEEE 24th Annual International Symposium on Personal, Indoor, and Mobile Radio Communications (PIMRC), London, pp. 876–880 (2013)
23. Xu, H., Zhou, Z.: Cognitive radio decision engine using hybrid binary particle swarm optimization. In: 2013 13th International Symposium on Communications and Information Technologies (ISCIT), Surat Thani, pp. 143–147 (2013)
24. Haldorai, A., Ramu, A., Chow, C.-O.: Editorial: Big Data innovation for sustainable cognitive computing. Mobile Netw. Appl. (2019)

25. Anandakumar, H., Umamaheswari, K.: Energy efficient network selection using 802.16g based gsm technology. J. Comput. Sci. **10**(5), 745–754 (2014)
26. Anandakumar, H., Umamaheswari, K.: Cooperative spectrum handovers in cognitive radio networks. In: EAI/Springer Innovations in Communication and Computing, pp. 47–63 (2018)
27. Anandakumar, H., Umamaheswari, K.: A bio-inspired swarm intelligence technique for social aware cognitive radio handovers. Comput. Electr. Eng. **71**, 925–937 (2018)
28. Xin, Y.: China. In: Dementia 3Ed, pp. 261–264 (2005). https://doi.org/10.1201/b13239-38
29. Cavalcanti, D., Schmitt, R., Soomro, A.: Achieving energy efficiency and Qos for low rate applications with 802.11e. In: IEEE Wireless Communications and Networking Conference, Kowloon, pp. 2143–2148 (2007)
30. Ying, C.L., Yonghong, Z., Peh, E.C.Y., Anh, T.H.: Sensing throughput tradeoff for cognitive radio networks. IEEE Trans. Wirel. Commun. **7**(4), 1326–1337 (2008)
31. Chandra, R., Mahajan, R., Moscibroda, T., Raghavendra, R., Bahl, P.: A case for adapting channel width in wireless networks. ACM SIGCOMM Comput. Commun. Rev. **38**(4), 135–142 (2008)
32. Shrivastava, R., Shrivastava, S.K., Shrivastava, A.K.: Study of microwave dielectric charecteristics of soil in North East Chattisgarh. J. Pure Appl. Indus. Phys. **8**(12), 214–218 (2018)
33. Monti, S., Soltanian, A., Zadeh, H.: Improved particle swarm optimization and applications to Hidden Markov Model and Ackley function. In: IEEE International Conference on Computational Intelligence for Measurement Systems and Applications (CIMSA) Proceedings, Ottawa, ON, Canada, pp. 1–4 (2010)
34. Song, C., Zhang, Q.: Intelligent dynamic spectrum access assisted by channel usage prediction. In: INFOCOM IEEE Conference on Computer Communications Workshops, San Diego, CA, pp. 1–6 (2010)
35. Tokel, T., Aktas, D.: A low-complexity transmission and scheduling scheme for WiMAX systems with base station cooperation. EURASIP Journal on Wireless Communications and Networking. **2010**(1), 527591 (2010)
36. Nachef, V., Patarin, J., Volte, E.: Generic attacks on classical feistel ciphers with internal permutations. In: Feistel Ciphers, pp. 75–94 (2017)
37. Liu, X., Zhang, C., Tan, X.: Double-threshold cooperative detection for cognitive radio based on weighing. Wirel. Commun. Mob. Comput. **14**(13), 1231–1243 (2013)

Chapter 4
Software Radio Architecture: A Mathematical Perspective

4.1 Introduction

In consideration to Software Radio Architecture (SRA) perspective, it is evident that communication devices due to the establishment of internet technology have enhanced the conveyance of the relevant data to the relative receivers. Due to the advanced engineering of software radio, devices necessitate encoding data that originate from a specific sources being driven into a particular electronic receiver. Resultantly, the transforming device needs to transit the internal data form into a designed waveform that is compatible with a definite radio frequency channel [1]. The radio channel defining the frequency influences the RF signals, forms noises and formulated the resultant time delays, which are a distortion due to RF signal replicas. This process inaccurately reversed into the neighboring receivers as indicated in Fig. 4.1. Considering this historic instance, software radio control necessitates future research, especially when analyzing the limited on and off power, noise threshold receivers, RF switch for manual selection of predefined channels and volume controls. The receivers and transmitters situated and operating in collection are composed of the radio node. On the other hand, the multi-nodes, multi-bands, multi-thread, and the multi-personalities capacities of software radio necessitate extensions from the mathematical framework.

4.2 SRA Functions

The advancements evident in technology have led to the development of new dimensions of software radio that are composed of a physical category of flexibility. The multi-band is speculated to compose of a single communication framework band, which indicates the channels indicated in Fig. 4.1 are characteristically

© Springer Nature Switzerland AG 2019
A. Haldorai, U. Kandaswamy, *Intelligent Spectrum Handovers in Cognitive Radio Networks*, EAI/Springer Innovations in Communication and Computing, https://doi.org/10.1007/978-3-030-15416-5_4

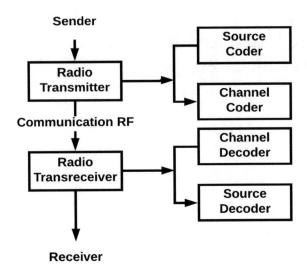

Fig. 4.1 Analogue RF Communication Channel

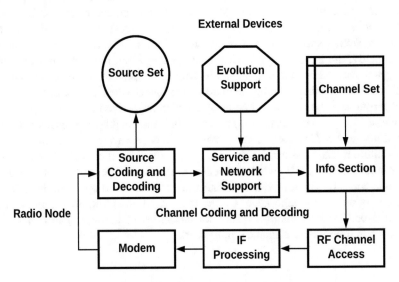

Fig. 4.2 Featured software radio model and communication scheme

generalized to a set of channel and communication scheme in Fig. 4.2. The sets defined in Fig. 4.2 include the channels that define the radio frequency. However, the nodes such as the individual communication scheme and base framework are portable radios that interconnect the cable and fiber, which are inclusive of the sets of channel. The radio encoding features are expandable during the programming of RF channel accessibility, intermediate radio frequency and processing. The conversion of RF and antennas are capable of spanning a collection of radio frequency bands, including the fibrotic interface, which consist of the radio frequency channels function accessibility. The intermediate RF processor includes more transformation of RF, filtering, beam forming, time/space diversity process, and other fundamental computation functions [2]. The multi-node radio delivers a collection of air interfaces and waveform that are illustrated fundamentally in radio frequency framework modulator and demodulator. When the software radio gradually advances, the joining controls enhance their complexities, which involve the autonomous choice of the nodes, bands, and information appearance. A particular RF function might appear single toned, for instance one band compared to a collection of bands and complicating joining controls. Responsive beam transformation allows more users, which also enhances the service quality.

4.3 Software Radio Architecture's Mathematical Perspective

The contribution given in this research attests that future research provided in Fig. 4.2 is featured with a mathematical approach to software radio architecture. This perspective constraints the definition of the play and plug module, which abstracts the data relative to the implementation of interfaces and radio functions. This module acknowledges the geometrical function and computation which the hardware and software features are composed of. The future mathematical approach focuses on the model of signal mapping in the software radio architecture framework [3]. This interface is denoted by its vertices evident in the topology spacing, which are illustrated as geometrical spaces with a set of theory axiom. In various interface sections, mappings embrace the presentation like arcs and edges. In this instance, edges are the sections which intersect two relative vertices. The radio frequency ASIC, for instance, is capable of transforming the signal located at the input apex to a base-band signal located at an output apex. On the other hand, the arc represents the map with features located at the vertices like the power that is dissipated through radio frequency ASIC transformation. The model is significant when

- Identifying higher levels of play and plug interfaces
- Predicting the performance of the control system
- Defining the models of reference illustrating the standardized scenario
- Deriving the principles of software radio architecture on the evolution of products

The multi-processor mathematical approach specifications of tasks centered on the topology spaces are applicable when proving significant features defining the shared memory multi-processor. The mathematical perspective illustrated in this contribution is parallel to these models. The model given in Fig. 4.3 indicates that software radio and hardware feature linking the interface links to the topology spacing. The sample included indicated a dual-receiver band composed of traditional to modern converters IF, the channels filter ASIC, voice codes, and the digital software features. Most of the arc stands for the signal and information movement from a single point of interface to an isochronous actual-moment stream [4]. Other

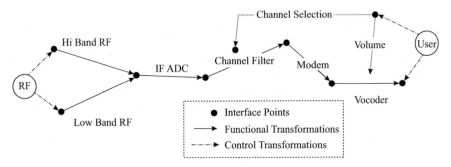

Fig. 4.3 Illustrative software radio topology framework

arcs stand for the exertion control for states (like the selection of channels) and a parameter which affects the entire stream.

Different forms of the arc's properties assume the overall dimensions of topological spaces. For example, the cross sections nodes that are situated in the IF-ADC composed of the channeling filters are composed of hardware figures defined by dimension evaluations.

4.3.1 SRA Purposes

The play and fitting architecture is inclusive of various rates that define two components. These components are:

- Consistency: The modulation structures are tasked with an obligation to secure the software radio architecture and its characteristics in the control framework. The structures need not to be composed of unexpected findings or effect exorbitant computation structures.
- Correspondence: The play and fitting modulations need to be effective and similar with the software radio status. This necessitates bending the plugs to hubs to determine the capabilities of the software radios.

The software radio architecture defines the modulation and structure whereby multiple segments are used to attain the evaluated administrations and capabilities in the configuration and limitation radio protocols. In that regard, the correspondence and consistency are the features of the play and fitting architecture, whereby the empirical framework for the play and plug architecture significantly induces the ideal features of a particular segmented arrangement in the formulated planned protocols. Since the open-completed status of the software radio innovations and administration, the software radio structure need to be extended in the protocol administration and the different unique protocol steps [5]. The basic analysis of computing features of the software radio segments at a single point formulates a phase aimed at determining the covered virtual machine structure. The software radio architecture protocols which assist in the play and attachment of the structure are made concurrently.

4.4 Software Defined Radio (SDR)

The Software Defined Radio (SDR) represents the telecommunication network whereby the components have inclusively been implemented in a traditional manner into system hardware. The example of implementation systems are the filters, mixers, detectors, demodulators, and modulators. The implementation is executed in these systems other than on various software means based on individual embedded and computer systems. Although the SDR concept is a traditional aspect, the quickly developing capacity on the modern electronic systems executed practically according to various processes had been considered in a theory form earlier on. The basic evaluation of the SDR framework consists of the individual computation tools that are composed of the voice cards, inclusive of the other traditional and modern converters, which precede some forms of radio frequency fronts and ends [6]. A considerate measure of the signal process is delivered on the overall duty processors instead of being executed on particular purpose hardware or the electric circuit. This form of software radio design is perceived to receive and transform the extensively unique software radio protocols, which are otherwise referred as the waveforms. Moreover, these protocols are majored on the degree of software utility [7]. The SDRs are characterized categorically for their fundamental use for the telecommunication and military services that necessitate the application of a greater dimension of the transforming software protocols in an actual timeframe. Moreover, the SDRs are also predictable by the various components such as the SDR forums that are commonly referred to as the Wireless Innovation Forums. These forums enable the users to embrace technology and become paramount innovators in SR communication. This aspect of technology together with the SDRs antenna is fundamental for the enhancement in cognitive radio networks [8].

The SDR is characteristically flexible with a principle purpose of avoiding the scarce spectrum access and assumption for the users who have user the traditional forms of radios in various manners [9], which include.

- The application of the ultra-broadband and spreading spectrum technique, which is fundamental for allowing various users and transmitter to transfer their data in a single location and frequency with minimal to no interference, are principally linked with a single or more error-determining techniques. These techniques are necessary for correction of any forms of mistakes that might be detected in a particular interface.
- The SDR antennas are adapted on the locked and direction frequency that enables the receivers to effectively reject any forms of interference in a particular direction. This allows the users to effectively determine the fainter transmission [10].

- Application of the cognitive radio network techniques. In every software radio dimension, there is possibility of measuring the spectrum that is in communication and utility with the cooperating software radios based on data transfer. This also enables the transmitters to avoid any mutual form of interference through the selection of any un-utilized frequency bands. Otherwise, every software radio interlinks with the geographical location and dataset that is necessary for the process of delivering information linked from the respective receivers. Moreover, this also aids in lowering the amount of transmitted data to a particular standardized minimum, including the reduction of near to far issues and the reduction of various interferences to the nearby signals [11].
- Evaluation of the networking meshed network which includes any forms of included overall radio capabilities and power reducing devices in a particular node. Every node principally transforms loud enough to enable the necessary information to hop to the closest nodes in a particular locus. Moreover, it also aids in the reduction of near to far issues and the diminishing to relative interference to one another.

4.4.1 SDR Software Architecture

The availability of the software radio initiative is much inspiring and a trending aspect of technology waited to enhance the future communication network. This system is fundamental for the emphasis of the wireless telecommunication systems that includes the enhancement of the third generation network, mobile phones, and the upcoming generational network, including the rural next generational networks. Because of its significant reprogram ability, flexibility, lowered costs of investment, scalability, software defined radio, and throughput, the business wireless telecommunication sector is now facing some significant issues as a result of the outstanding issues from the rapidly evolving linked layers of protocols and standards. Moreover, the availability of the variable wireless technology in various nations inhibit the deployments of the worldwide roaming systems and issues in bringing out unique features and services. This advancement comes due to the global availability of licensed users and devices.

The software radio initiatives promise a significant advancement in the telecommunication sector since it targets at solving critical issues through an implementation of the software radio function. The software modulation that runs on the nonspecific hardware systems implements various principles, which are evident in the software radio frameworks. These systems are capable of assuming various characteristics based on different software modulations under utility. Moreover, these modulations which apply unique features and services can be retrieved from the upper air dimension to a particular desired handset. The form of flexibility that is implemented by the software defined radio framework aids in the mitigation of issues as a result of the differing issues and standards connected to a leveled deployment of various features and services. The enhancement of communication frameworks centered on software radio systems necessitates effective analysis of the architecture behind the SDR alongside the necessary software and hardware [12]. Whereas the software

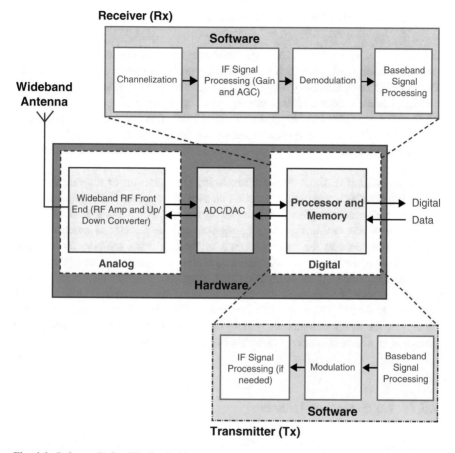

Fig. 4.4 Software Defined Radio Architecture

defined radio initiative delivers a significant measure of merits, presently there is no business based software radio that appears in the telecommunication market. The main reason for this rationale is its demand for a number of computation resources required to affect the performance of different SDR functions. The fundamental aspect of formulating the software defined radio is the inclusion of the DAC and ADC features, which are a division between the digital and analogue domains.

Therefore, the processing signals may be transmitted using software. The architecture behind SDR may be considered based on an analysis of both software and hardware. The DAC and ADC placement evaluates the digital and analogue domains, whereby the analogue domain is formed by the hardware and the digital one is formed by software. In that manner, these domains are fundamental for the separation of both the digital and the analogue signal processors which are commonly referred to as the digital accessing points. These points are critical for the application of various software radio functions which are delivered through the software process. Figure 4.4 included indicates both the software and the hardware architecture aspect of the software defined radio structure.

4.5   Calculation Features of Function Elements

A fundamental segment of SRA is the structuring of the status defining the connection of function findings, which are critical for analyzing the computing elements. The aspect of compensability defining the features of the overall topology spacing is a unique literature aspect. Homeomorphism indicates the topology safeguarding the maps, which are formulated to produce various homeomorphisms into a single scope and guide of different topological elements. Based on a single topology point, the software radio area composed of the overall contribution is composed as shown in Fig. 4.5.

The evaluation of the architecture includes a comparison of relevant findings, which implies that counting the values and relevant reactions is fundamental. The software radio includes the designing of systems that are composed of timing necessities which enable the users to suggest various limitations controlling the topology structures in the radios. Moreover, these systems are also fundamental when setting up status and the software modulation arrangement exceeds the limited resources anticipated. The assumption that forces the users to utilize the limited resources comes from the desire of formulating the exceeding radio capacity arrangement whereby various limited resources might be put into maximum utility. Moreover, the evaluation and analysis of these capabilities need to be limited in a manner that will have less impact on the in-ordinate resources.

4.5.1   Computation Models

The ideation behind the software radio proportion model originates from the FPGA and ASIC which may be ascribed as the computing capability of various unlicensed processors. These processors are displayed as a collection of different Von-Neuman machines. Each machine and system that is encoded in the software in the Von-Neuman processors maybe illustrated in connection to the firm multi-faced status and calculability status of the regular accessible machine. This machine

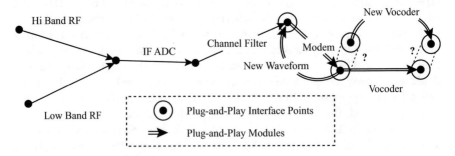

Fig. 4.5 Topological Plug-Play Interface

includes the stats machine, result clustering, dataset, and the inner licensing [13]. Moreover, it also includes the capabilities of storing data and loads, which applies the models of back-handing and immediate locations with the capabilities of increasing the model of the memory. The software radio machine moreover has been illustrated effectively on the recursive manners, inclusive of the Turing mechanism and figuring model. Fundamental hypothesis elements incorporate the capabilities to imitate the polynomial timeframe capabilities which necessitate an exponential timeframe of the Turing machine. Reflecting on the radio machine model, the software architecture may be detected as ISA which is a subjective processor. A hypothesis which is delineated under uses the recursive capability, including the radio machine model in setting the upper limitations in formulating the resources, which is also a virtual status for the throughputs.

4.5.1.1 Traditional Recursive Element

The methodology of the recursive element in the process of computing includes the categorization of the "N" elements and the analogue numbers whereby "N" is a composition of the enclosing feature that illustrates the categorization elements. The traditional recursive usually encloses the classification functions, which implies that the "y" element equates to 1 and "h" xn equates to "h" xn. In this manner, "f" actualizes the execution status of the "g" and "h" relative to the "y" valuations. Thus, the pattern of the recursive form equates to the valuation when programming contraction that develops a form of computability. In an event that "f" acts in a manner that bears its own definition, it is fundamental to pull down the stacks based on the successive findings. Nevertheless, because the valuation of "y" is initially exceeding the value "0," the "y" element usually diminishes in single invocations that eliminate the stacks and procedures that extensively eliminates any forms of an overflow. In that regard, it is evident that the allocation of the stack overflow in a certain ISA actualizes the stack finite based on the overflow when the top bounds are developed. Thus, the finding explaining the bounds is composed of the software radio applications in the architectural theory.

4.5.2 General Recursive Element

In the traditional recursive model, the iterative loop is never included. Moreover, these loops are critical in the comprehension of the software radio architecture. The base functions of the loops are a bound minimalization which is expressed by allowing "g" being "N^{n+1}." This expression indicates the traditional recursive models that are referred as the crude recursive functions shown in Eq. (4.1) below.

$$f : N^{n+1} \rightarrow N : f(x,y) = \mu z \{z < y\} \{g(x,z) = 0\} \tag{4.1}$$

When element "μz" is taken as "$f(x, y)$," and if it is a minimum of "z" and does not exceed "y," it is then assumed as "$g(x, y)$" equating to zero. The classification of capacities diminishes considering the crude recursion, piece, and the scarce minimalization taken to be minimal. The Ackerman capacity is illustrated as the recursive resources that are effective at any confirmed capability due to the structure of the software recursive categorization. This capacity is never considered as the crude recursion, although it can be regarded as the absolute capability. Moreover, the finite parameter defining the Ackerman's capability that is evaluated ahead of a given timeframe enables one to decide whether the capability will be greater than the upper resources by using effective stages that evaluates the capability. A crude recursive capability may incite the crude recursive developed from the laying of the capabilities.

A form of scarce recursive capabilities equates to the radio machine program that ends with a finite figure of the steps characterized in terms of limited resources. This is the minimalization whereby the composite capacity is formed. The proof framework is inclusive of a collection of radio machines direction that is composed of scarce basic capabilities and formulated findings. Each interconnected radio machine guide and successions are basic recursive and scarce. The similarity aspect of the radio machine processor is devoted to any forms of scarce recursive that has the capabilities to execute its toile of time measurement, which might be limited. A user may evaluate an upper execution timeframe from every radio machine class guide based on a certain ISA [14]. This bound of execution timeframe is due to the quantity of the theory venture characterized by multiple timeframe execution in a class role.

The fundamental significance of the recursive element is to induce the software radio to fuse isochronous mainstreams whereby there are firm timings window among which certain capabilities have to be executed with an end ambition for the users to convey potentially and appropriately for the stability of the framework. The modulations composed of structures are not mandated to make use of the assets limited on the software architecture and experiments; however, this modulation affects in various status. Various basis recursive capacities are capable of ensuring two rationales: the first is the completion in a certain window and the second one is to ensuring it is critically processed in advance and un-equipped in planning and meeting the window.

4.5.3   The Partial Recursive Function

The scarce basic recursive capacity does not illustrate the various programming enhancements evident in software radio networks. Significantly, there are circles that detect the status whereby the recursive will end. This status and condition will not take place, which means that the radio machine program will be set to circle without a design time length. The software and equipment forms, which hang firms critically, scan the status which might partially take place and are also computationally comparative to resultant circles. The flow and Ebb analysis that evaluate the available calculations signify the resultant problems in the guidance and succession will end in a finite step figures. The software radio source codes (the SpeakEasy I) fundamentally feature the overall distant utility [15, 16]. One critical example of this source code is the firm sitting for enough solidarity for recursion. The feature and procedure usually dispatch the circles on the bit operative systems, which principally expends the infinite resources that convey them to various software radio applications. Moreover, the remaining forms of infinite circles are computable and proportional to the fraction recursive capabilities. The structure of computation is known as the unbounded minimalization as indicated in Eq. (4.2) shown below.

$$g : N^{n+1} \rightarrow N, f(x_n, y) = \mu y \left(g(x_n, y) = 0 \right) \tag{4.2}$$

Equation (4.2) shown above shows that "y" represents the minimal figure, whereas $g\,(x_n, y)$ equates to 0. The questioned users denoted by μ shall resultantly maintain on their argumentation "y" while still testing "g" without any bound forms. The limited recursive capabilities represent the most minimal capacity arrangement that is enclosed in minimalization. However, not all the similar minimalization that expend the finite resources represent the level at which the calculations are done in advance. The unbounded forms of minimalization affect the resources to a point when "g" aspect is fulfilled. In most instances, "g" aspects may render unfulfilled. For example, the geological aspect of the bearer and circle situated at the enclosed flag which will dismiss the transmission of the signal due to the standardized radio frequency will render to be infinity. This aspect will progressively utilize the resources to the junction when every user applies the ventures that constrain the utility of the limited resources. The partially confirmed recursive capabilities are illustrated and similar to those of the Turing, radio machine display and the post formulation of the frequency. The capacity value principally is standardizing based on the users' realization on the model processing. The capacity of software radio is evaluated with the circles and reciprocals, i.e., the execution that applies the programming frameworks may never be insured in the application of scarce computing resources. Equation (4.3) below indicates the proof blueprint that necessitates the construction of minimalization.

$$(g\,(x_n, y) = 0) \tag{4.3}$$

An individual may depict signs of equality of until, though, with the loops for the unrestrained minimalization. In the long run, this can be unequal and not well defined. Therefore the equal RAM protocols may be in a position of getting through the resources of the unbound algorithm, "crashing" the entire framework. The set instructions of the programmable RAM's processor usually incorporate a set piece of instructions in which the unbounded loops might be structured. On the other hand, there is a comprise of realistic methods in which most radio engineers together with the real-time programmers who have devised make sure that there is system steadiness despite the fact that there is ordinary utilization of the contracts. A specific individual might be in a position of allocating specific time limits for a particular search operation. In the long run, a time out scenario may result to a fault which involves extreme consumption of time. Hence such programming practices will be changing to the unbounded minimalization. On one hand, the time restrictions should incorporate an arithmetical outcome of rival for various resources using operations that are of higher precedence.

On the other hand, the radio applications are very separate centered from the general purpose algorithm applications since there comprise of vast timing set limits induced within the emission, creation, reception, conversion, processing, user interface of the radio signals, and network inter-operations. Additionally, the time limits frequently correspond to state machines and message sequential charts. The timeframe limitation is significantly considered on the set-up and control, which includes the video and voice isochronous delivery, small package connection, multimedia operation and association controlling systems. In that regard, the set-up includes a large time-framed domain application [17]. The constraints mainly involve the timing, with resource bounds which can easily be allotted to specific software and hardware features. Such bound hampers the reasonable (both hardware and software) to the tightly bounded algorithmic constructs without deprived loss of radio functionality. An official task for a particular bound is mainly the background for the theory based on the plug and play software radio steadiness which is shown as follows. Theory 3 (Normal viability of the partial bounded recursion): on threaded single processor software radio operation can be amalgamated with the use of RAM procedures succession equal to the resource inhibited subsets which comprise prejudiced recursive functions. With a single processing software radio function, it can be synthesized within the RAM procedures series which are equal to the resource limited subsets within the incomplete recursive function.

Constructions outline (The local partial bounded recursion): This evidence is set based on theory 1 and 2. The while and unlimited loops together with their go-to equal do not assure the termination which should be precluded. A single construct depicted from the conservative enactment is thus allotted a higher number of iterations (or rather equal, or a higher processing time).

4.6   The Interface Topologies Within the Play and Plug Modules

Limitation based on the algorithmic structures for the modules can ensure bound resources with the use of play and plug. Are there any limitation based on the interface structure within such modules which can additionally promote the play-plug? The most accurate reply is based on the necessity of complementary aspects defining the topology modules.

4.6.1   Topological Gaps

Definition: Essentially, a topological space, symbolized by, X, Ox is the x set comprising of an Ox subset which is always an open subset. The sets always include X together with an exposed set($\varnothing$) which has been surrounded within restricted intersections and known unions. Contrarily, the composition of a topology is a family subset in which Ox constitutes of a geometric model and an algebraic structure. On the other hand, amid the topological gap of the geometric elements within the crossing points is always signified within the radio modules. For an analogue point of view, it can always be structured solely as a subset functional module amid R, in which it adheres to certain limitations [7]. Therefore such restrictions confer to lower and upper transmission bandwidth, power, with opposite channel interference. There is always a guarantee of numerous waveforms when there is an analogue interface, though most dictatorial bodies and other systems limit the waveforms to being a subset of credible waveforms. Since the gaps comprise of disconcerted signals within the metric space, the distance is always assumed alongside the waveforms, thus the topological models inside the analogue edge are analogue waveforms alongside the time domain (a) with n frequency domain (b).

4.6.2   Limited Interface Topologies

There are always an open number of subsets within a collection of elements |X|. When $|X| = M$, X will be $2^{(2M-2)}$ since the topological number would have convinced the topological number since it's a dual exponential. Based on the previous results it can be clearly noticed that not all of the topologies are in agreement with the closure of the linked and the finite junctions as compelled by the topological gaps. A huge number of likely topologies are compelled based on certain finite interface topologies in an accurate manner. Meaning: With set B being the center basis of Ox when the clients Ox have linked up and related with B. In this segment, the lower set rationalized with the Ox maybe provoked using the unions. Based on hardware perspective, both lining pins are always profound to being the crossing point known as $xi \in X$ subset $\left\{ X = \bigcup_N xi, \varnothing \right\}$ will be for the base Ox. On the other hand,

$Y = \left\{ \bigcup_{N-1} xi, \{x1\}, \varnothing \right\}$, which consists of three subsets that is $\{x1\}$, $\varnothing$, wholly with the oneness of other interfaces might pin them making them operate in a similar manner. Thus the middle comprising of a vast number of subsets is $\varnothing$, $\{xi\}\}$, while the $N + 1$ comprises of the sets which are inclusive of every single element of X and an open element in which it is fully handled as a single set of elements. Within an event in which most pins are always necessitated for essential correlations, thus, $\{X\}$ will be single subset of X that will consummate with the linkage as the only element among the interface topology.

Hence, $Ox = \{\{X\}\}$ alongside with other topologies will thus be a fixed set as necessitated by the set inputs of X. Generally, all this can be known as rigid topology. On the contrary, this cannot be the topological gaps because it denies an empty space within it. An empty set does not constitute of valid users alongside the interface, thus not working when the pins are present. Upon ejection of the connectors, the whole system goes down and thus the vacant sets are in a position of linking with the other sets within the topological sets. Therefore, such a thought is highly applicable to most software. This idea is highly related to software. For the formation of a particular function (), a specific call can be subjected rapidly consisting of a certain argument RF and the bandwidth. Both of them are essential, thus $Oy = \{\{Y\}\}$ and $Y = \{RF, W\}$. In the long run, this can be un-estimated unlike other interfaces. Therefore using a tagged API of equation $W = 30$ or RF 859, both of them can be stated among one or more spaces consisting both $\{RF\}$, $\{W\}$, $X = \{RF, W\}$, and $\varnothing$: The unaccomplished sets within the incorporated topologies despite the fact that the interface operates without any contentions are also provided. Consequently, with $Ox = \{X, \{RF\}.\{W\}, \varnothing\}$, the alignment of the X subsets and the power sets alongside are known as the topological gaps. Thus the power set topology is now the discrete topology, or on the other hand it is referred as flexible API topology. Such flexible interfaces can be used in spite of the fact that they exhibit certain default characteristic comprising missing contentions.

4.6.3   Capacity Call Parameter Topologies

A better comprehension of the geometric structures can be fully understood using this concept. Simplex: a simplex basically entails an alignment of different focuses within a topological space which are adjacent in some relations, for instance using a connection R. Greater measurement simplexes actualize by bringing down the dimensions simplexes. Thus the simplexes can be inserted within the required space. Complex: a complex is an association of many simplexes which involves the association of major low dimensions of simplicities for a particular simplex. The Q-linked: simple structures which share the $q + 1$ face are linked. As shown in Fig. 4.6, vertices which are A, B, and C comprise a two-dimensional simplex majorly adjacent and also linked through focusing within the plane. Thus each line segment adjoining the vertices is inclusive of a one-dimensional simplex, pointed out by the 3 vertices as shown in Fig. 4.6.

In the model more than one expression could be BW and RF using MathCAD broaden documentation. The unequivocal commence can be maintained by the earth to yield a bumble result "out of range" for conflicts that misuse the start depiction.

Fig. 4.6 Simplex Contain
dual Simplexes

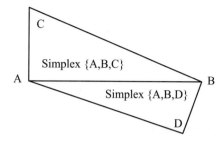

There is a limitation of 200 kHz exchange ability to the RF's apportioned to GSM, say, 925 to 960 MHz. A simplex defines the set-supplement subspace for which the 200 kHz exchange speed choice is mapped to a gripe condition. The degree of subspaces can be changed while the radio center point is in the field; the master community has a "future check" interface. Changing the field "limit" is definite as the interface constraints must be defined so that the control computations can test and control the restrictions.

4.6.4   Attachment and Play Interface Geometry

Using decisions made in the midst of arrangement, headway, fuse, and test may propel an interface to a point in the simplified complex of interfaces. Every zone and extent of a normal interface includes a doled out point in its simplified complex. The physical interface subspaces must change as a component of the hardware in which the organization is passed on. A reasonable interface subspace similarly may need to change as a segment of the product modules congaed to pass on the organizations [18]. Being totally extensible, connection and play modules must be combined intensely. Controllably, the control algorithm must have a technique for differentiating the extent of one limit with the space of the close to choose if the limits are immaculate.

4.6.5   Extensible Capabilities

Restrictions may be addressed as level marks, for instance, the sort of video remotely organizing interface in the ITU H.320 Proposal. An upside of marks is straightforwardness. Capacity marks are defined in the midst of the standards setting process and revealed in the substance of the standards. One bother of this approach is the possible blunder of the substance of the standard. Countering this consistent issue in correspondence standardization, ETSI broadcasted vernaculars that makes the formal specification of the ITU Z.100 specification and depiction tongue the institutionalizing explanation of the standard, with substance giving amplification and elucidation.

4.6.6   Architecture Partition

The primary objective of connection and play programming radio engineering is to surrender enrolling resources from as of not long ago undefined hardware modules to help as of recently undefined programming modules for as of not long ago undefined organizations. Taking into account engineering examination, the computational and topological properties begin to perceive the deductively based portions of a product radio system. A source code of a for the most part dispersed military programming radio, SpeakEasy I, is considered. On the other hand, a course of action of virtual machines discovered that both broadens the Slam rise to direction set model and oblige the computational geometry, yielding a layered virtual machine reference show for the product radio.

4.6.7   SpeakEasy I

A developed code has some strong features, for instance, progressing execution and correct treatment of timing contrasts between radio frameworks. Due to the fact that an Ada was directed, the progressing official is the Ada run-time bit. The Ada modules can be viewed as programming objects. There was consolidation of the databases, channels, and masters. Databases store personalities, alter parameters, long chunks of accumulated code, and instructive files to be stacked into a character at run time. Channels are consultations around which modes are dealt with. Channel systems are reinforced by a couple of Ada packages that carry out the structure level components of RF control, modem taking care of and related inter-networking activities [19]. It presents its personality on these resources for realizing a mode. It by then screens the general state of the planning string that passes on the related organizations. In SpeakEasy I, cut down measurement modules realize the person-alities of the channels.

4.6.8   The Hardware-Specific Partition

In this section, the software and hardware specifically state the lowest dimension sections of the software radio frameworks. Based on the topological aspect, the processor tools and equipments interface the proper conditions of the data that is required for the registration. The alignment of data state which easily accessed within one cycle within the underlying conditions of processors trademark is simple. Therefore in this section a processor with a speed of about 20–32 bits registers, harbor the points, and other important registered user should also be able to aspire for a state condition of about 2(32–20) which is defined as 101,926 = 0.358846241 M2. It will thus reduce the level of the processor momentary trademark simplex as stated in this section. On the other hand, the less registered clients within the group will produce minimal simplex. Routes within the simplexes induced partitions, linked simplexes which are q-associated with each other through clock alignment. Due time, the sequential follow-up of the RAM directions is in compliance with a series of succession of simplexes, thus the linking up is simplified. During 1 s, a processor comprising of about 100 MHZ always takes control of the clock exclusive of those having about $(10192) \wedge (108)/087822$ M 5 states. Additionally, with a software radio which is disseminated and a multi-processor and interconnectors will comprise of extra linked multi-processors. For a particular set of clock cycle, an edge simplex inter-crossed (e.g., a rigid bolt processor of 1 s within time $k = 3$).

The Interrupted Service Routine (ISR) Topological loops: This particular software operates in a manner that it rigidly couples various equipment and other associated processors. Thus, a unique place which is the ISR, for example, the equipments may be destroyed as the interruptions proceed, pushing the processors, evaluating the interface with a condition, and also setting a dispatch. Within a linear framework, certain protocols may take into account about 10 to 100 instructions. With Windows NT the data layout can increase in size to about 100 directions and more.

Topology substrate surface: This section incorporates the use of software precisely in order to adjust various states and consequently comprising of the topological circles. Thus a processor consisting of a collection of tools and other subordinate modules can be exhibited using machines which wholly cover the entire subset, inclusive of the, ISA, RAM bit. Thus the subdivided segments may comprise of known sets of RAM guidelines which are for the inactive states. Therefore, the inactive RAM within the idle space exhibits a subjective alignment that is easily conceivable. Therefore, when restricting the amount of RAM separation for the necessitated segments increases the RAM limits thus higher portion of the radio segments. Thus the simplex of the RAM sections comprises of a known subspace linked to the modes in which they can be accessed as a through part.

4.6.9   Software Radio Infrastructure Topology

Provided that there is a vast portion of software used, this section mentions some sectors which have the capability of designing, structuring, and taking control of natural assets in order they can be legit. Therefore such capacities are taking into account of the structures recommended within the ITU's X900. About 900 open distributed preparations show such a scenario. Additionally they both exhibit structures which link to the software radio applications, such as the recurrence circulatory networks. Therefore the main source code takes charge in a variety of ways which include timing, flag flows, and other accompanying data. Consequently, this also takes into account various necessitated isochronous distributions of video, audio, and other streaming services. Thus the control flow channels affect the messages transmitted alongside various articles within certain frameworks. Various errors, semaphores that look over linked assets, and the distributions, always get to the conventions which highly control the flow messages over the whole systems. The main source capabilities introduce the system and also take control of various legit and other physical ports, distribute messages, and also execute remote method callings specifically referred as the RPC.

4.6.10   Radio State Machines

A vast number of software radio assets are fully controlled and subdued by the state machines. The main functions of the state machines are often regulated, broadcasted, and also acquire channel systems recuperation events. Therefore the main composition of the state machines includes: waiting back for a certain reaction, aligning together a waveform, and waiting back for an instantiation. Hence the curves are thus stated based on certain conditions which are in alignment with a particular state then to the next event and certain activities executed based on specific progress. The control structures can be performed solely as a software character comprising of a vast number of gaps and with the appended methods as indicated in Fig. 4.7.

Minimum level advanced machines can be in a position of tracking the condition of dynamic crypto adjustments which convey simple and innovated linked blurs. Consequently the SpeakEasy I conditions are in compliance with the state machine during timing and error circumstances. The AGC and the squelch are equally stable amid every corner with each transition between the lower phases within the state machines. In such a manner, such state machines are in a position of scheduling various routine protocols, such as the AGC and the protocols initiated channel conditions, i.e., going over the blurs. Additionally, the states are in a position of

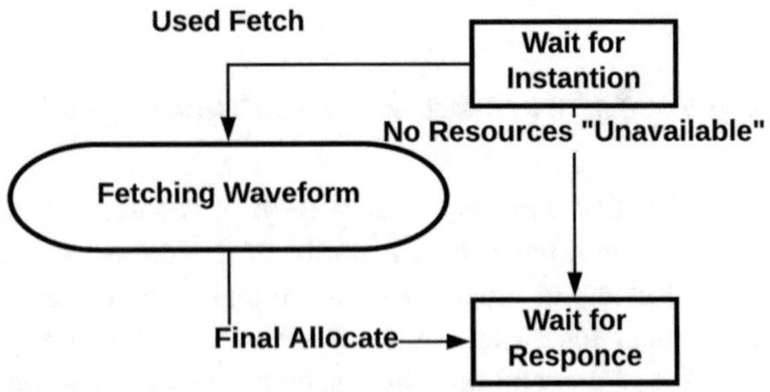

Fig. 4.7 Radio infrastructure states machine

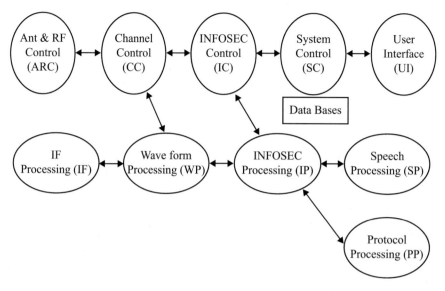

Fig. 4.8 Channel-agent software features

depicting the failure within the different nodes. It takes into account various uncertainties, the distribution failures, and the deprivation of framework assets as indicated in Fig. 4.8.

4.6.11 Channel Agents

There are characteristics of SpeakEasy I and other software radios which have been delinked. Such an unusual condition can be termed as channels specialists. Consequently every particular software feature has been assigned a software radio role. Therefore, the specialists who best perform the process as outlined in Fig. 4.8 are responsible of ruling the entire network, while the ones in the lower channel take part in actualizing the traffic and other associated administration events. Consequently, every associate within this abnormal state software comprises of subordinate articles, which reveal dimensions that are crude within dissimilar capacities such as bit decisions, filters, and many more. There is utilization of the software radio within the convention and other discourse handling techniques such as TCP/IP, Mobile IP, ATM, and many more. Thus the inter-networking within the wire foundation consists of a variety of similar rigid pre-stated software features.

4.6.12   Dispersed Layered Virtual Machine Referencing Model

Earlier sections mentioned about the depicted software apparatus which also use the techno-advanced software radio systems. To this point, consider the fact that all topological gaps are stated based on certain simple identical RAM system, together with progressively reduced machine based learning software. With a high level of distributed systems, virtual machines used for speaking to various chains is basically a technique of conveying messages to the particular set of command system. Therefore each layer is a set of command based on the virtual machine within the subordinate layer. In the long run, the virtual machines might cause holding back of the usefulness of various subtleties and other confined modernized inventions which can be present within the section based on a particular set of layers unlike other layers which are highly avoided. Thus the software radio architecture signified in this case is fully shown in Fig. 4.9. Due time the restrictions of both DSP and FPGA maintain the current position by developing and also being prepared of aiding certain virtual machines which are in turn being at a moderate level. The management provides within the virtual based machines which are free of tools performance and the best layer virtual machines [20]. Hence the DSP and the FPGA equipped performance is processed based on the software source providers. There can be a provision of essential groups within the software infrastructural sector. The layers are applicable in various sections using the augmented data that penetrated in the interfaces. The states machine composing the radio infrastructure and various

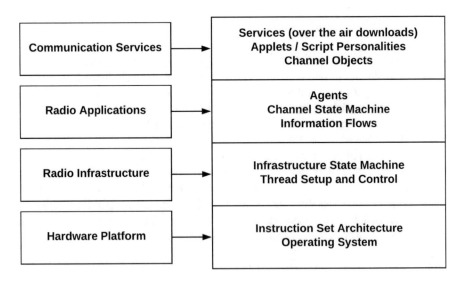

Fig. 4.9 Distributed and layered virtual machine

applications in the radio layers physically diminish in the SDL domain. Other waveforms and vendor would then be composed of the middle form category and layer, whereby the interface in the components can be presented in UML or the IDL.

4.7 Summary

In conclusion, the software radio architecture follows a mathematics approach, which composes fundamental geometrical and computation elements that shape the level of the play and plug. This field of research has now been realized in the worldwide market since it aids users to have a greater acknowledgement and understanding towards sophisticated mathematical elements related to software radio architecture. However, this form of architecture needs to accelerate its application to becoming cost-friendly for play and plug services that are aspired by developers worldwide. The mainstream architectural principle aiding the acceleration is the bounding recursive module, which includes the construction of software radio model that hosts the plug and play module. The relative modules will be evaluated based on the different inputs which will consume different predictable framework applications despite the idea of the software containing some faults. However, these faults are less likely to affect the application and delivery of any data, audio, and multimedia services. Moreover, there are fewer chances of the faults to affect the framework or even crashing it compared to unconstrained module. Another principle is the explicit and extensive topology interface, which necessitates defining the software interfaces via an explicit model that underlines the topology spacing.

References

1. Mitola, J.: Software radio architecture: a mathematical perspective. IEEE J. Select. Areas Commun. **17**(4), 514–538 (1999)
2. Ozone, M., Hiramatsu, T., Hirase, K., Iizuka, K., Tomisawa, S.: Reconfigurable processor based on ALU array architecture for software radio. Int. J. High Perform. Syst. Archit. **3**(1), 33 (2011)
3. Anandakumar, H., Umamaheswari, K.: Supervised machine learning techniques in cognitive radio networks during cooperative spectrum handovers. Clust. Comput. **20**(2), 1505–1515 (2017)
4. Haldorai, A., Ramu, A.: Cognitive social mining applications in data analytics and forensics. In: Advances in Social Networking and Online Communities (2019)
5. Anandakumar, H., Umamaheswari, K.: Energy efficient network selection using 802.16g based GSM technology. J. Comput. Sci. **10**(5), 745–754 (2014)
6. Li, S., Singhoff, F., Rubini, S., Bourdellès, M.: Scheduling analysis of tasks constrained by TDMA: application to software radio protocols. J. Syst. Archit. **76**, 58–75 (2017)
7. Mitola, J.: The software radio architecture. IEEE Commun. Mag. **33**(5), 26–38 (1995)
8. Zhang, L.: Software architecture evaluation. J. Softw. **19**(6), 1328–1339 (2008)
9. Lingaiah, D.: Software radio: a modern approach to radio engineering [book review]. IEEE Softw. **20**(4), 86–95 (2003)

10. Li, R., Dou, Y., Zhou, J., Deng, L., Wang, S.: CuSora: real-time software radio using multi-core graphics processing unit. J. Syst. Archit. **60**(3), 280–292 (2014)
11. Anandakumar, H., Arulmurugan, R., Onn, C.C.: Computational intelligence and sustainable systems. In: EAI/Springer Innovations in Communication and Computing (2019)
12. Arteaga, A.: Architecture of a spectrum monitoring system using software-defined radio. Sistemas y Telemática. **10**(23), 83 (2012)
13. Kalyanaraman, S., braasch, M.: GPS adaptive array phase compensation using software radio architecture. Navigation. **57**(1), 53–68 (2010)
14. Anandakumar, H., Umamaheswari, K.: Cooperative spectrum handovers in cognitive radio networks. In: EAI/Springer Innovations in Communication and Computing, pp. 47–63 (2018)
15. Anandakumar, H., Umamaheswari, K.: A bio-inspired swarm intelligence technique for social aware cognitive radio handovers. Comput. Electr. Eng. **71**, 925–937 (2018)
16. Sahukar, L., Madhavi, L.: Frequency domain based digital down conversion architecture for software defined radio and cognitive radio. Int. J. Eng. Technol. **7**(216), 88 (2018)
17. Qing, L., Kai, C., Ying-yong, L.: FPGA software architecture for software defined radio. Proc. Eng. **29**, 2133–2139 (2012)
18. Savic, D., Pavlovic, B., Sunjevaric, M.: Software: based radio architecture. Vojnotehnicki glasnik. **48**(1), 48–54 (2000)
19. Subramanian, N.: Software architecture interference—an important non-functional requirement for software ecosystems. Int. J. Softw. Archit. **1**(1), 15–16 (2010)
20. Suganya, M., Anandakumar, H.: Handover based spectrum allocation in cognitive radio networks. In: 2013 International Conference on Green Computing, Communication and Conservation of Energy (ICGCE), Chennai, pp. 215–219 (2013)

Chapter 5
Distributed Algorithms for Learning and Cognitive Medium

5.1 Introduction

Distributed algorithms for learning and cognitive medium have undergone an extensive research over the past few decades. This area of research is obliged to provide a resolution to challenges that had not been mitigated during the convectional wireless networks. A fundamental challenge in the field of distributed algorithm is the attainment of coexistence of heterogeneous subscribers who have an access to spectrum parts. In cognitive media access, there are two fundamental categories of transforming users: these are the Viz, (Primary Users) PUs with the priority to spectrum access and the (Secondary Users) SUs with opportunity to transform in the PUs are idle. The cognitive ones are the SUs who have the capabilities of sensing the spectrum access to enhance the detection of PUs transmission [1]. Additionally, as a result of hardware and resource constraint, these users can be able to visualize a spectrum section provided at any moment. This article considers a slotted cognitive media access framework that permits every SU to access and sense an orthogonal channel in every transformation slot as indicated in Fig. 5.1. Considering the sense constraint, it is fundamental for SUs to choose the frameworks characterized by an extreme mean capability. These frameworks are also featured with minimal likelihood of being occupied by the PUs. Practically, the accessibility of cognitive media channels and statistics is previously unknown to the SUs due to the rationale that necessitates sensing of the medium access prior the transmission. The rationale also permits the sense of decisions utilized in media channel learning of available statistics. Resultantly, the application of formulated media channel accessibility leads to the designation of channel access principles that reduce the throughput transformation. A formulation of efficient algorithms that can be proved to apply machine learning is a critical focus of this article. The algorithms necessitate efficiency in both medial channel accessing and machine learning.

© Springer Nature Switzerland AG 2019
A. Haldorai, U. Kandaswamy, *Intelligent Spectrum Handovers in Cognitive Radio Networks*, EAI/Springer Innovations in Communication and Computing, https://doi.org/10.1007/978-3-030-15416-5_5

Fig. 5.1 CR networks composing of $U = 4$; SUs and $C = 5$ channels

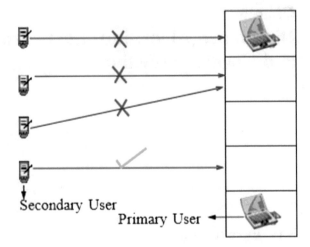

Secondary User

Primary User

For learning algorithm and cognitive medium access, two fundamental evaluation criteria, which are regret bounds and convergence, are applied. Considering the context of this paper, estimations are necessary to converge to actual availability of channel statistics representing the amount of accessible sense decisions, which might proceed to infinity. A firm criterion is represented by the learning algorithm regret that measures the overall convergence speed. Considering the context illustrated in this research, the regret represents the loss of secondary throughputs that arises due to learning in reference to identifying the channel statistics effectively [2]. Therefore, it is critical to denote the learning algorithm, which contains minimal regrets. These regrets represent a fine assumption of learning algorithm performance compared to an average throughput considering the implication of sublinear regret to offer optimal average throughput. Moreover, this research considers a distributed system that is stipulated by minimal data exchange or initial policy agreement with SUs. Resultantly, this raises more challenges like the throughput loss arising from the SUs' collision and the completion evident among the SUs who effect to access medium channels composed of a higher accessibility. Imperatively, distributed channel access principles need to overcome the problems identified in the research since the cognitive access principles and distributed learning encounter regret as a result of learning the unidentified medium channel accessibility and distributed learning access.

5.2   Cognitive Medium Contribution

The primary focus of this article is on distributed algorithms for learning and cognitive medium. The paper proposes two effective learning and access principles for manifold SUs in a cognitive network. Moreover, this research delivers a performance assurance of the principles in consideration to regret. As an overall assumption, this research proves that the proposed algorithms attain an orderly optimum regret while also attained closely order optimum regret, whereby the orders are denoted by the amount of transformation slots.

According to Fig. 5.1, the SU is disallowed to transmit when the available framework is utilized by the PUs. When more than one SU transforms in the existing available framework, all the transformation renders ineffective.

The first arrangement we propose denotes that the all out number of auxiliary clients in the framework is known while our second strategy loosens up this necessity [3]. Our second strategy likewise fuses estimation of the quantity of auxiliary clients, notwithstanding learning of the channel availability and structuring conveyed get to rules. We give limits on all out lament experienced by the auxiliary clients under self-play, i.e., when actualized at all the optional clients. For the first approach, we demonstrate that the lament is logarithmic, i.e., O (log n), where n is the quantity of transmission spaces. For the second arrangement, the lament becomes marginally quicker than logarithmic, i.e., O ($f(n)$log n), where we can pick any capacity $f(n)$ fulfilling $f(n) \rightarrow \infty$, as $n \rightarrow \infty$. Thus, we give execution certifications to the proposed appropriated learning and access strategies.

In this manner, our first approach (which requires learning of the quantity of optional clients) accomplishes arrange ideal lament. The impacts of the quantity of optional clients and the quantity of channels on lament are additionally unequivocally described and verified through recreations. To the best of our insight, the investigation abuse exchange off for learning, joined with the participation rivalry exchange offs among various clients for dispersed medium access have not been sufficiently analyzed in the writing previously (see Section I-B for a dialog). Our investigation in this paper furnishes critical designing experiences towards managing learning, rivalry, and participation in handy intellectual frameworks. Comment: We take note of a portion of the deficiencies of our methodology. The i.i.d. demonstrates for essential transmissions are in fact optimistic and by and by, a Markovian model might be progressively proper. In any case, the i.i.d. show is a decent estimate if the schedule openings for transmissions are long and additionally the essential traffic is exceptionally bursty. Additionally, the i.i.d. show isn't essential towards determining lament limits for our proposed plans. Extensions of the established multi-furnished scoundrel issue to a Markovian model are considered. On a basic level, our outcomes on appropriated learning and access can be comparatively stretched out to a Markovian channel show; yet this involves increasingly complex estimators and standards for assessing the exploration and exploitation trade-offs of various channels and is a subject of enthusiasm for future examination.

A few outcomes on the multi-outfitted outlaw issue will be utilized and summed up to contemplate our concern. The nitty gritty exchange on multi-equipped highwaymen represents an intellectual medium access, which is a subject of broad research. The association between subjective medium access and the multi-outfitted l issue provides an anxious brigand definition. Under this plan, index ability is set up, the Whittle's record for divert determination is acquired in shut shape, and the identicalness between the nearsighted approach and the Whittle's file is set up. Notwithstanding, this work accepts known channel accessibility insights and does not consider contending auxiliary clients [4]. This research considers distribution of two clients to two channels under Markovian channel show utilizing a mostly perceptible Markov choice process

(POMDP) structure. The utilization of impact criticism data for learning, and spatial heterogeneity in range openings were examined.

In any case, the distinction from our work expects that the accessibility measurements (change probabilities) of the channels are known to the optional clients while we think about learning of obscure channel insights. Our works consider incorporated access plots rather than appropriated access here; our works also consider access through data trade and concentrate the ideal decision of the measure of data to be traded given the expense of arrangement. Our works also consider access under Q-learning for two clients and two channels where clients can detect both the channels at the same time. The work talks about an amusement theoretic way to deal with subjective medium access. In [5], learning in clog amusements through multiplicative updates is considered and assembly to feebly stable equilibrium (which decreases to the unadulterated Nash balance for practically all recreations) is demonstrated. Be that as it may, the work expects fixed costs (or comparably remunerates) rather than arbitrary rewards here, and that the players can completely watch the activities of different players. As of late, the research comprehends about combinatorial marauders, where an increasingly broad model of various (obscure) channel availabilities is accepted for various optional clients and a coordinating calculation is proposed for together dispensing clients to channels. The calculation is ensured to have logarithmic lament as for number of transmission openings and polynomial stockpiling prerequisites. A decentralized execution of the proposed calculation is proposed; however regardless it requires data trade and coordination among the clients.

Interestingly, we propose calculations which expel this necessity though in a progressively prohibitive setting. We first planned the issue of decentralized learning and access for numerous auxiliary clients. We thought about two situations: one where there is introductory regular data among the auxiliary clients as pre-designated positions, and the other where no such data is accessible. In this paper, we examine the appropriated arrangement in detail and demonstrate that it has logarithmic lament. What's more, we likewise consider the situation when the quantity of optional clients is obscure, and give limits on lament in this situation [6]. We proposed a group of circulated learning and access strategies known as time-division decent amount (TDFS) and demonstrated logarithmic lament for these approaches. They built up a lower bound on the development rate of framework lament for a general class of consistently great decentralized policies. The TDFS arrangements can consolidate any request ideal single-player approach while our work here depends on the single-client strategy.

5.3   System Model and Formulation

Given any combine functions $[f(n), g(n), f(n) = O(g(n))]$ there is a relatively constant "c" this implies that $f(n) \leq cg(n)$ that represent all the $n \geq n\ O$ in the constant $n\ O \in N$. Likewise, the $f(n) = \Omega(g(n))$ is evident when there is an existence of constant "c," which indicates that $f(n) \geq c'g(n)$ stipulated is in every $n \geq n\ O$ given in every constant $n\ O \in N$ and the $f(n) = \varnothing(g(n))$ when $f(n) = \Omega(g(n))$ and $f(n) = O(g(n))$. Moreover, $f(n) = O(g(n)$ if $f_g^n(n) \to 0$ with $f(n)\omega(g(n)$ if $f(n)/g(n) \to \infty$

representing $n \to \infty$. Thus the calculations are in reference to the U, which is accorded the maximum vector entry and the most effective channels while the rest applies to U as the worst medium channel. Letting $\sigma \, (T : \mu)$ to indicate the greatest Tth entry in μ, we shall abbreviate $T * \, = \sigma \, (T : \mu)$ which makes it effective for notation. Considering the worst notation, let $D \, (\mu1 \, \mu2)$ to be equal to $D \, B(\mu1) : B(\mu2)$ which signals the Kullback Leibler distance over the Bernoulli principle $B(\mu1)$ and $B \, (\mu2)$ by letting $\Delta(1, 2) = \mu1 - \mu2$.

5.3.1 Cognitive Channel Models and Sensing

To evaluate this rationale, let $U \geq 1$ which signifies the number of SUs and $C \geq U$ indicate the number of orthogonal frameworks in the transformation channels with set slot measurement. In every channel "i" and slot "k" the PU transforms "i.i.d" which have the probability of $1 - \mu i > 0$. This implies that $Wi(k)$ denotes the actual variable indicator when the cognitive channel is available. Thus, it is assumed to be $W_i(k) \overset{\text{i.i.d.}}{\sim} B(\mu i)$.

$$W_i(k) - \begin{cases} 0, & \text{channel } i \text{ occupied in slot } k \\ 1, & \text{o.w} \end{cases}$$

For vector "μ," the availability of the mean is composed of the "μi" availability representing all the forms of channels. This implies that μ equals $(\mu1$ and $\mu C)$ which signifies $\mu i \in (0, 1)$ being distinct as μ. This assumption is previously unidentified to SUs hence studied independently over a specific time frame with the application of sense decisions eliminating any forms of data stipulated in the specific users. In this analysis, it is assumed that the preliminary transformation is errorless to both the secondary and the primary users. Thus, letting $T_{ij} \, (k)$ to indicate the amount of slots whereby the "i" channels are sensitive in "k" slot, the SUs $j \in U$ chooses a single channel $i \in U$ on sensing and therefore it obtained a specific value denoted by $Wi \, (k)$ that shows whether the channels are available. The user denoted by j records the ongoing sense decision stipulated in "k" slots in all the medium channels "C." Resultantly, it is assumed that when the model collision whereby two users or more transform equal channels, no transformation of channels will take place. In the end of the slots denoted by "k" every "j" user attains acknowledgement denoted by $Zj \, (k)$ which indicate if the transformation in "kth" slots is attained. Generally, the principle shown by "j" in $k + 1 - th$ slot indicated by $\rho \, (UC \, i = 1Xk \, (ij)ZK \, (j)$ is centered on an initial feedback and sense assumption.

5.3.2 Policy Regret

In reference to the model indicated above, the designing of the ρ policy is fundamental since it maximizes the speculated figure of effective transmissions of all the
SUs that are tantamount to minimal interference constraints for all the PUs. Letting
S $(n : \mu, p)$ to signify the speculated overall number of effective transformations
following the "n" slot, U will signify the number of all the SUs and ρ. An ultimate
instance that denotes the accessibility μ statistics is identified as a major agent and
priori orthogonal identifies the SUs as the most effective channels, then the speculated number of effective transmissions following the slots is indicated by

$$S^*(n; \mu, U) := n \sum_{j=1}^{u} \mu(j^*), \tag{5.1}$$

whereby j^* represents the greatest jth entry in the μ.

Algorithm 1 Single User Policy ρ^1 ($g(n)$) in [7].

Input: $\{\bar{X}_i(n)\}_{i=1,...,C}$: Sample-mean availabilities after n rounds, $g(i; n)$: statistic based on $\bar{X}_{i,j}(n)$, $\sigma(T; g(n))$: index of Tth highest entry in $g(n)$.
Init: Sense in each channel once, $n \leftarrow C$
Loop: $n \leftarrow n + 1$
$Curr\_Sel \leftarrow$ channel corresponding to highest entry in $g(n)$ for sensing. If free, transmit.

Evidently, S^* $(n : \mu, U, p)$ in every policy denoted by p conforms to finite n. Thus,
there is necessity to reduce the regret in cognitive access and learning to conform to
Eq. (5.2) given below.

$$R(n; \mu, U, p) := S^*(n; \mu U) - S(n; \mu, U, p) > 0. \tag{5.2}$$

Moreover, it is necessary to reduce the regret indicated in $\mu \in (0, 1)$ C that denotes
all the distinct elements. Through the corporation of collision frameworks assumption model, there is minimal avoidance of mechanisms, which implies that the
throughput considering the ρ is indicated as shown in Eq. (5.3) below.

$$S(n; \mu, U, p) = \sum_{i=1}^{o} \sum_{j=1}^{u} \mu(i) EV_{ij}(n), \tag{5.3}$$

whereby V_{ij} (n) shows the amount of "n" slots and the "j" user represents the major
user sensing the "i" channel. In this instance, the regret is evaluated as

$$R(n; p) = \sum_{k=1}^{U} n\mu(k^*) - \sum_{i=1}^{o} \sum_{j=1}^{u} \mu(i) EV_{ij}(n). \tag{5.4}$$

5.4 SU Instance Findings

The analysis exemplified in this research recaps on the regrets in specified instances of one SU denoted by $U = 1$ with the various users characterized by a central cognitive access and learning by appealing with the effective multi-armed and an effective finding in the banding process.

5.4.1 One SU U = 1

With only one SU, the issue on policy finding characterized by minimal regret diminishes to a multi-armed bandit progression. Initially, the schemes on multi-armed progression characterized with asymptotic algorithmic regressed were centered on a higher confidence bound on unidentified available channel. From that instance, the simple schemes compute an index and statistic of every arm and channel [8]. This is denoted as the G-statistic considered on effective mean and the amount of slots, which is the point at which the aims are sensed. The arm indicated by the upper index is chosen in every slot stipulated in this research. Thus, it is assumed that ρ in the first algorithm is denoted by ρ 1 (g (n), whereby g (n) represents the score vector designated to the framework due to transmission "n" slot. The example mean base ρ signals an index in every framework "i" and users "j" at a particular timeframe is shown by

$$g_j^{\text{MEAN}}(i; n) := X_{ij}(T_{ij}(n)) + \sqrt{\frac{2 \log n}{T_{ij}(n)^2}} \tag{5.5}$$

From the above Eq. (5.5), ($T_{ij}(n)$) represents the slot number, whereby "j" chooses channels "i" significant for sensing and Eq. (5.6) below represents the actual channel availability "i" sensed by "j."

$$X_{ij}(T_{ij}(n)) := \sum_{k=1}^{T_{ij}(n)} \frac{X_{ij}(k)}{T_{ij}(n)} \tag{5.6}$$

The exploitation and exploration trade-offs over the sense channels characterized by a superior predictive accessibility minimize the significant throughput which senses various medium channels to determine enhance availability estimates. The example mean term is in correspondence to the exploitation, whereby the next term denoted by $(T_{ij}(n))$ is in correspondence with the relative exploration due to its penalized framework that is less sensed. When the medium channels are fully sensed θ (log n) count of times, the exploration definition renders less significant on the G-statistic of the cognitive channel, which causes the exploitation term to dominate. Resultantly, it favors channel sensing composed on the greatest mean sample, whereby the regret centered on statistics is algorithmic in various finite slots denoted by "n" but lacks an optimum scale constant. This also leads to an optimum scale regret constant represented as follows.

$$gj^{\text{OPT}}(i;n) := X_{ij}(T_{ij}(n)) + \text{mean}\lfloor\sqrt{\frac{\log n}{2T_{ij}(n)}}, 1\rfloor \tag{5.7}$$

The principles applied in this article are centered on g^{mean} framework and statistic, which signifies a recap of results that indicate an algorithmic regret to learning the most effective framework. In this instance, there is standardized best ρ denoted by

$$R(n;\mu, U, p) = o(n^{\alpha}), \quad \forall \alpha > o, \quad \mu \in (0,1)^{c} \tag{5.8}$$

As for the theory 1 (algorithmic regret that characterizes U equal to 1), any standardized best policy applies to the speculated timeframe in sub-optimum channel i which equates to 1^{*} and satisfies Eq. (5.9) indicated below.

$$\lim_{n \to 100} P\left[(T_{ij}1(n) \geq \frac{(1-\epsilon)\log n}{D(\mu_{ij}\mu1^{*})}; \mu\right] = 1, \tag{5.9}$$

From Eq. (5.9), 1^{*} shows the framework composed of the most effective form of accessibility. Thus, the regret is shown as

$$\liminf_{n\to\infty} \frac{R(n;\mu, 1, p)}{\log n} \geq \sum_{i\in 1} \frac{\Delta(1^{*}, i)}{\Delta(\mu i, \mu 1^{*})}, \tag{5.10}$$

Therefore, the regret denoted by the statistic g^{OPT} attains the bound shown above

$$\lim_{n\to\infty} \frac{R(n;\mu, 1, p^{1}(gj^{\text{mean}}))}{\log n} = \sum_{i\in 1} \frac{\Delta(1^{*}, i)}{\Delta(\mu i, \mu 1^{*})}, \tag{5.11}$$

Algorithm 2 Centralized Learning Policy ρ^{CENT} in [9].

Input: $\chi^n := \bigcup_{j=1}^{U} \bigcup_{i=1}^{C} X_{i,j}^n$: Channel availability alter n slots, $g(n)$: statistic based on χ^n, $\sigma(T; g(n))$: index of Tth highest entry in $g(n)$.

Init: Sense in each channel once, $n \leftarrow C$

Loop: $n \leftarrow n + 1$

$Curr\_Sel \leftarrow$ channels with U-best entries in $g(n)$. If free, transmit.

The regret denoted by the statistic g^{MEAN} satisfies

$$R\left(n; \mu, 1, p^1(gj^{\text{mean}})\right) \leq \sum_{i \neq 1*} \Delta(1^*, 1) \left[\frac{8 \log n}{\Delta(f^*, i)^2} + 1 + \frac{\pi^2}{3} \right] \qquad (5.12)$$

5.5   Central Cognitive Access and Learning for Several Users

This title considers several SUs in a central medium access policy which is characterized by a joint access and learning on a centralized agent who represents all the other users. To make the calculation of the regret to be minimal, the centralized policies allocate the "U" to an orthogonal framework which eliminates any forms of collision. This assumption is executed by letting p^{CENT} with $X(k)$ to be equal to U and j to be equal to 1 and UCi to be 1, $X(k_{ij})$ represents a central principle centered on sense variations of the users. The principle representing a central learning is standardized to a uniform policy as indicated in the second algorithm [10].

As for theory 2, which is the regret on a central policy p^{CENT}, any standardized effective central principle p^{CENT} ensures the actualized timeframe in U signifies the worst medium channels that "i" satisfies.

$$\lim_{n \to \infty} P\left[\sum_{j=1}^{u} T_{ij}(n) \geq \frac{(1-c)\log n}{D(\mu_{ij}, \mu u^*)}; \mu \right] = 1 \qquad (5.13)$$

From the above Eq. (5.13), U^* represents the Uth effective accessibility. Therefore, the regret is shown as

$$\liminf_{n \to \infty} \frac{R(n; \mu, 1, p^{\text{CENT}})}{\log n} \geq \sum_{i \in U} \frac{\Delta(U^*, i)}{D(\mu_{ij}, \mu u^*)} \qquad (5.14)$$

The schemes shown in the second algorithm centered on the g^{OPT} attain the bound shown above

$$\lim_{n\to\infty} \frac{R\left(n;\mu,1,p^{\text{CENT}}\left(g^{\text{OPT}}\right)\right)}{\log n} \geq \sum_{i\in U} \frac{\Delta(U^*,i)}{D\left(\mu_{ij},\mu u^*\right)} \tag{5.15}$$

The schemes shown in the second algorithm centered on the g^{MEAN} assure that n is greater than 0

$$R\left(n;\mu,1,p^{\text{CENT}}\left(g^{\text{MEAN}}\right)\right) \leq \sum_{M=1}^{U}\sum_{I\in U}\sum_{k=1}^{U}\frac{\Delta(m^*,i)}{U}\left[\frac{8\log n}{\Delta(m^*,i)^2}+1+\frac{\pi^2}{3}\right]. \tag{5.16}$$

5.6   Preliminaries: Regret Bounds

Satisfied with a classic finding on multi-armed bandit, there is necessity to formulate the distributed learning and access principles. This necessitates the provision of standardized bounds on the regrets that represent any forms of cognitive access and distributed learning policy denoted by ρ. The first proposition is the upper and lower regret bounds, whereby the regret in distributed policy is given as [11]

$$R(n;p) \geq \sum_{j=1}^{u}\sum_{i\in U}\Delta(U^{*},i)ET_{ij}(n), \tag{5.17}$$

$$R(n;p) \leq \mu(1^{*})\left[\sum_{j=1}^{u}\sum_{i\in U}ET_{ij}(n)+EM(n)\right] \tag{5.18}$$

where $T_{ij}(n)$ is the quantity of openings where client j chooses channel I for detecting, $M(n)$ is the quantity of impacts looked by the clients in the U-best diverts in n spaces, $\Delta ij = \mu(i) - \mu(1^{*})$ is the most astounding mean accessibility. In the resulting segments, we propose conveyed learning and access strategies and give lament certifications to the approaches utilizing the upper bound. The lower bound can be utilized to determine bring down headed on lament for any consistently great arrangement. The first term speaks to the lost transmission openings because of choice of U-most exceedingly terrible channels (with lower mean availabilities), while the second term speaks to execution misfortune because of impacts among the clients in the U-best channels [12]. The first term decouples among the diverse clients and can be broken down exclusively through the peripheral conveyances of the g-measurements at the clients. This thusly, can be broken down by controlling the established outcomes on multi-armed crooks. Then again, the second term, including crashes in the U-best channels, requires the joint appropriation of the g-insights at various clients which are associated factors [13]. This is recalcitrant to examine straightforwardly and we create systems to tie this term.

ρ^{RAND}: Distributed Learning and Access. This paper provides the ρ^{RAND}: policy in the third algorithm. Before portraying this arrangement, we mention some straightforward objective facts. In the event that every client actualized the single-client arrangement in Algorithm 1, it would result in crashes, since every one of the clients focuses on the best channel [9]. At the point when there are different clients and there is no immediate correspondence among them, the clients need to randomize direct access so as to maintain a strategic distance from crashes [14]. In the meantime, getting to the U-most exceedingly awful channels should be maintained a strategic distance from since they add to lament. Subsequently, clients can stay away from impacts by randomizing access over the U-best channels, in light of their evaluations of the channel positions. Nonetheless, if the clients randomize in each opening, there is a finite likelihood of crashes in each space and this outcomes in a straight development of disappointment with the quantity of schedule vacancies [7]. Subsequently, the clients need to join to a crash free configuration to guarantee that the lament is logarithmic. In Algorithm 3, there is versatile randomization dependent on criticism with respect to the past transmission. Every client randomizes just if there is a crash in the past opening; generally, the recently created irregular position for the client is held. The estimation for the channel positions is through the g-measurement, on lines like the single-client case.

Algorithm 3 Policy $\rho^{RAND}(U, C\ g_j(n))$ for each user j under U users. C channels and statistic $g_j(n)$.

Input: $\{\bar{X}_i(n)\}_{i=1,\ldots,C}$: Sample-mean availabilities at user j after n rounds, $g_j(i: n)$: statistic based on $\bar{X}_{i,j}(n)$, $\sigma(T; g_j(n))$: index of Tth highest entry in $g_j(n)$.
$\zeta_j(i; n)$: indicator of collision at nth slot at channel i
Init: Sense in each channel once, $n \leftarrow C$, $Curr\_Rank \leftarrow 1$, $\zeta_j(i; m) \leftarrow 0$
Loop: $n \leftarrow n + 1$
if $\zeta_j(Curr\_Sel; n - 1) = 1$ **then**
Draw a new $Curr\_Rank \sim \mathrm{Unif}(U)$
end if
Select channel for sensing, If free, transmit.
$Curr\_Sel \leftarrow \sigma(Curr\_Rank; \mathbf{g}_j(n))$.
If collision $\zeta_j(Curr\_Sel; m) \leftarrow 1$, **Else** 0.

5.6.1 The Regret Bound Within the ρRAND

As noticed, it can be easily viewed that the ρRAND strategy makes sure that all system clients are allotted [15] orthogonally based on U-best channels since the transmission slot numbers have become expanse. Therefore, the regret bound of the ρRAND has not yet shown to be precise are assured in the section beneath in the research paper. Initially, the upper logarithmic upper limits within the number of slots utilized are a single client is provided in each U-worst channel [5]. Consequently, the initial term within the bound on regret is known as logarithmic. The duration of time utilized within the U-worst channels (Lemma 1): Beneath the ρRAND system of Algorithm

3, entire time used by a particular client given as $j = 1, U$ in a particular $i \in -$ worst channel is shown as

$$\mathbb{E}|Tij(n)| \leq \sum_{k-1}^{U} \left[\frac{8 \log n}{\Delta(i,k^*)2} + 1 + \frac{x2}{3} \right]. \tag{5.19}$$

The evidence provided is similar to the evidence provided by the Theorem 2 as exhibited in [16]. Thus our main attention should be drawn in investigating the collision numbers $M(n)$ within the U-best channels. First, the outcome will be provided based on the prospected collision numbers within the idyllic circumstance in which every single client has concise acquaintance of the number of channel statistics availed μ. For this scenario, every client endeavor is attempting to get to the orthogonal (collision free) setting through consistent randomizing more than the U-best channels [16]. Thus the stochastic process in this case has shown in a finite condition of the Markov chain. On the other hand, with a condition within the Markov chain related to the system setting of the U number for the indistinguishable clients within the U channel number [17].

Markov chain state number is the count of the compositions of U, provided by $2U - 1 U$. The orthogonal system setting compares to the retaining state. Considering other states, comprising many clients or rather no clients within any of the channels, then the changing likelihood of any state within the Markov chain will be consistent (inclusive of self-transition likelihood). Moreover, for any particular state comprising prospective channels that have precisely one client, then there will be only changes to states comprising of about one client within the channel while the transition likelihood will be the same [18]. Considering $Y(U, U)$ as the maximum absorption time within the Markov chain commencing from the first distribution channel, the accompanying outcome of Lemma 2 (# collision within an accurate awareness) will be: prospected collision number within the ρRAND plot as shown in Algorithm 3, presuming that every single client has ideal knowledge about the mean channel availabilities μ, as shown in Eq. (5.20) below

$$\mathbb{E}\left[M(n); \rho RAND\left(U, \mathsf{C}, \mu\right)\right] \leq \leq \left[\left(\frac{2U - 1}{U} \right)^{UE[Y(U,U)]} - 1 \right]. \tag{5.20}$$

Therefore, the presented outcomes expressed show that there is a finite number of an anticipated collision which is bounded by UE[$Y(U, U)$] with concise knowledge of μ. Conversely, recalling the data from the previous sections, it is exhibited that there is no known collision under ideal information of μ present to the designated positions. Thus, UE[$Y(U, U)$] signifies a bound for the extra regret as a result of the inadequacy of concise communication amid both clients in analyzing both their positions. Therefore, utilizing the outcomes of the Lemma 2 used in evaluating the collision numbers within the dispersed learning of the obscure availabilities μ as shown: in any event when the clients are in a position of learning the precise order of dissimilar channels comprising only the logarithmic regrets, therefore an additional finite prospected collision number will happen prior getting to the orthogonal system setting. Stating T′ (n; ρRAND) as the slot number with one from the top, U approximated positions of the channels with various clients being wrong within the ρRAND strategy. In the section below, there is proof which explains the prospected logarithmic values of the number of transmissions. Wrong alignment of the g-statistics (Lemma 3): Below the ρRAND method based on Algorithm 3,

$$\mathbb{E}[Tt(n;\rho RAND)] \leq U \sum_{a=1}^{U} \sum_{b=a+1}^{G} \left[\frac{8\log n}{\Delta(a^*,b^*)2} + 1 + \frac{x2}{3} \right]. \tag{5.21}$$

An upper bound is provided based on the collision numbers $M(n)$ within the U-best channels through slotting in the presented outcomes $E[T'(n)]$, thus the outcome of the total slot number $E[Ti,j]$ utilized within the U-worst channels amid the Lemma 1 while the collision average number $UE[Y(U,U)]$ within clear comprehension of μ in Lemma 2. Thus according to theory 3 (Comprising the logarithmic collision number within the ρRAND)

Thus, there exist the logarithmic numbers of the prospected collisions prior each client settling along the orthogonal channels. Consolidating the outcomes together with that of Lemma 1 which include the total number of slots utilized within the U-worst channels will also be logarithmic, then straight away the primary result for the research paper would be obtained generalizing both the regret within dispersed learning circumstances and access would be logarithmic. Logarithmic regret under ρRAND (Theory 4): The ρRAND rule (U, C, gMEAN) in Algorithm 3 comprises of Θ (log n) regret.

Subsequently, it is demonstrated that the conveyed learning and the channel access amid various secondary clients can be conceivable with the logarithmic regret with no clear correspondence between the clients. Therefore this infers that the number of lost chances of fruitful transmissions for all secondary clients will be logarithmic amid the number of transmissions, which shows that it is negligible in an event where there is high number of transmissions. So far great attention has been paid on structuring plans that augment frameworks and also by community throughput. Precisely, let's talk about the equality of a single client within the ρRAND. Because ρRAND is not familiar with any client, with respect that individual clients have equal likelihood of being aligned in one of the U-best channels while encountering the logarithmic regret when doing so.

5.7 Distributed Learning and Access Within the Unknown Number of Clients

Currently it is presumed that the prospected number of secondary client is not yet identified, since it is necessary for the execution of the ρRAND strategy. Therefore, practically speaking, it will entail the initial statements from one of the secondary clients signifying their presence within the cognitive networking systems. In any case, within the conveyed setting with no data sharing amid the clients, the declaration might tend to be difficult. In this segment considering the postulated circumstance, with the number of clients U not yet identified (although fixed within the entire duration of the transmissions and $U \leq C$, channel numbers). For this situation the policy should carefully approximate the number of secondary clients within the entire system notwithstanding learning channels availability data and structuring channels in getting access to the rules reliant on the collision criticism. You should note down that in an event when the policy presumes the bad case scene when $U = C$, therefore the regret will develop linearly because the U — worst channel will be picked severally for detection.

5.7.1   Explanation of the ρEST Strategy

According to Algorithm 4, it clearly shows the ρEST strategy. The strategy fuses two eligible functions considering each transmission slot, viz., implementation of the ρRAND strategy as shown in Algorithm 3, centered on the number of client's estimation bU, and brings to date the prospected estimate bU considering the collision numbers encountered by a single client. Thus the updating process depends in the possibility that there is an underestimation of U for all clients ($b\ Uj < U$ for all clients j); collisions essentially develop while the collision number acts as criteria of increasing bU. This is due to the fact that following an extensive learning process; every client would have vividly become familiar of the genuine positions of the channels while focusing the similar set of channels. Nonetheless, in any event where there is underestimation, thus the number of clients will surpass the channel number prospected by the client [19]. Thus in any event when there is a collision clients accumulate, which is effective for testing an increase in b.

Algorithm 4 Policy ρ^{EST} $(n, C\ g_j(m), \zeta)$ for each user j under n transmission slots (horizon length), C channels, statistic $g_j(m)$, and threshold functions ζ.

---

1. **Input:** $\{\bar{X}_i(n)\}_{i=1,\dots,C}$: Sample-mean availabilities at user j, $g_j(i; n)$: statistic based on $\bar{X}_{i,j}(n)$, $\sigma(T; g_j(n))$: index of Tth highest entry in $g_j(n)$. $\zeta_j(i; n)$: indicator of collision at nth slot at channel i U: current estimate of the number of users.

---

2. Init: Sense each channel once, $m \leftarrow C$, $Curr\_Rank \leftarrow 1$, $\hat{U} \leftarrow 1$, $\zeta_j(i; m) \leftarrow 0$ for all $i = 1,\dots, C$

3. Loop: $m \leftarrow m + 1$, stop when $m = n$.

4. **If** $\zeta_j(Curr\_Sel; m - 1) = 1$ **then** Draw a new **end if** Select channel for sensing. If free, transmit. $Curr\_Sel \leftarrow \sigma(Curr\_Rank; g_j(m))$

5. $\zeta_j(Curr\_Sel; m) \leftarrow 1$ if collision, 0 o.w.

---

6. $\displaystyle\sum_{a=1}^{m} \sum_{k=1}^{\hat{U}} \zeta_j\Big(\sigma\big(k; g_j(m)\big); a\Big) > \big(\xi(n; \hat{U})\big)$ **then** $\hat{U} \leftarrow \hat{U} + 1$, $\zeta_j(i; a) \leftarrow 0$, $i = 1,\dots C$, $a = 1$,

$\dots, m$. **end if**

Denotes the collision count used by ρRAND policy as:

$$\Phi kj(m) := \sum_{a=1}^{m} \sum_{b=1}^{k} \xi j(\sigma(b; gj(m)); a).$$

(5.22)

which is the overall collision count encountered by a single client j currently (until the mth transmission slot) within the top bUj channels, since the channel position is approximated with the aid of the g-statistics. Thus the collision number is evaluated alongside the threshold $\xi(n; b\ Uj)$, representing a function of length 6 horizons and the present estimate of bUj. Therefore, upon surpassing the threshold limit, bUj is increased, while the sample collision accumulated will be considered redundant (drawing them back to 0).

5.7.2   Regret Bound Under ρEST Value

In this section, there is an analysis of the regret bound under the ρEST strategy, with the regret denoted as shown on Eq. (5.3). Consenting to the highest threshold of the consecutive collisions number within the ρEST is shown:

$$\xi^*(n; U) := \max_{k=1\ U} \xi(n : k)$$

(5.23)

Proving the ρEST strategy comprises of $O(\xi * (n; U))$ during the time when $\xi * (n; U) = \omega(\log n)$, while n will be the transmission slot number. Therefore, the evidence of the regret bound within the ρEST strategy comprise two essential segments: The bound regret is proven in an event when the client over-approximation is U. This is maintained through making sure the through the selection of a suitable threshold $\xi(n)$ in evaluating the number of collisions extracted in the feedback segment.

Conditioned regret: The first outcome is initially presented. Explaining an (effective event) $C(n; U)$ with no client over-approximating U within the ρEST as shown below

$$C(n; U) := \left\{ \bigcap_{j=1}^{U} UjEST(n) \leq U \right\}.$$

(5.24)

Therefore, the conditioned regret value within $C(n; U)$ is shown as $R(n; \mu, U, \rho$ EST) $\mid C(n; U)$, provided as

$$n \sum_{k=1}^{U} \mu(k^*) - \sum_{i=1}^{C} \sum_{j=1}^{U} \mu(i) \mathbb{E}\left[Vij(n) | C(n; U)\right],$$

With $Vi, j(n)$ being the number of times in which each client j has possession of channel i. correspondingly, considering uncertain outcomes $E[Ti, j(n) | C(n;$

U)] together with the number of collisions within the U-best channel system is provided as $E[M(n)| C(n; U)]$. The regret state is thus exhibited within $C(n; U)$ and thus being $O(\max(\xi * (n; U), \log n))$.

Conditioned regret (Lemma 4): Upon the enactment of the U secondary clients in the pEST strategy, having all the $i \in U -$ worst channels with each client $j = 1, \ldots, U$

$$\mathbb{E}\left[ij(n)|\mathbb{C}(n)\right] \leq \sum_{k=1}^{U} \left[\frac{8 \log n}{\Delta(i, k^*)2} + 1 + \frac{\pi 2}{3}\right]. \tag{5.25}$$

Therefore the conditioned outcome based on the collision count $M(n)$ within the $U -$ best channels complies with Eq. (5.26) below

$$\mathbb{E}\left[M(n)|\mathbb{C}(n; U)\right] \leq U \sum_{k=1}^{U} \xi(n; k) \leq U2\xi^*(n; U). \tag{5.26}$$

Depicting from (5.15), it is shown that $R(n) | C(n; U)$ is $O(\max(\xi * (n; U), \log n))$ is for a particular $n \in N$. The over-approximation likelihood: It can be evidenced that no clients over-approximates 7 within the pEST strategy, for example, the likelihood of event $C(n; U)$ as shown in (5.22) nears 1 as shown, $n \to \infty$, during the event when the threshold limit $\xi(n; b\ U)$ utilized for analysis of the number of collisions is picked concisely (Algorithm 4, line 6). Insignificantly, $\xi(n; 1) = 1$ is set due to the fact that one collision is sufficient enough in designating that there are more clients. ξ is depicted in an event where there are other $k > 1$ satisfying

$$\xi(n; k) = \omega(\log n), \quad \forall k > 1. \tag{5.27}$$

It is thus evidenced that the above set measures make sure that the over-approximation do not happen at all. Remember, the $T'(n; \rho EST)$ is always the slot number comprising the top U-approximated positions of each channel within various client is incorrect within the pEST strategy. Therefore, it is shown that $E[T'(n)]$ will be $O(\log n)$. Duration of time utilized within the mistaken approximations (Lemma 5): Prospected slot number with any of the top U-approximated positions of the client channels will be mistaken within the pEST strategy satisfying Eq. (5.28) below

$$\mathbb{E}[Tt(n)] \leq U \sum_{a=1}^{U} \sum_{b=a+1}^{C} \left[\frac{8 \log n}{\Delta(a^*, b^*)2} + 1 + \frac{\pi 2}{3}\right]. \tag{5.28}$$

Noting the explanation of $Y(U, U)$ preceding the earlier segment, the greatest time to its assimilation commencing the first allocation of the finite Markov chain will be in line with the dissimilar clients system setting with the absorption condition being in line with the free collision system setting. Therefore, including the first explanation to $Y(U, k)$ being the absorbing time within the new fangled Markov chain, with the condition gap will be the system setting of the U clients as shown in the

k channels, while the changeable likelihood is stated within the same line. You should note that $Y(U, k)$ is nearly finite in an event when $k \geq U$ while ∞ or else.

Therefore the highest value of the number of collisions is bound as $\Phi k, j(m)$ within the ρEST strategy utilizing $T'(m)$ while the overall time duration utilized with the mistaken channel approximation, while $Y(U, k)$, the absorbing time duration with the Markov chain st $\leq$ symbolizes the stochastic arrangement for the two variables.

Proposal 2: Greatest number of collision within every client within the ρEST accomplishes the strategy below

$$\max_{j=1, \dots, U} \Phi k, \quad j(m) \overset{st}{\leq} (Tt(m) + 1)\gamma(U, k), \quad \forall m \in \mathbb{N}. \tag{5.29}$$

It can be verified that the likelihood of the highest approximation will be 0 asymptotically. With no over-estimations within the ρEST (Lemma 6): A threshold operation that satisfies the occurrence $C(n; U)$ also satisfies that

$$\lim_{n \to \infty} P\big[C(n; U)\big] = 1 \tag{5.30}$$

The primary outcome based on this segment is that ρEST comprises fairly a greater logarithmic regret asymptotically which is reliant on the threshold function $\xi * (n; U)$. The asymptotic regret within the ρEST (Theory 5): While ξ as the threshold function is agreeable with conditions, then the strategy ρEST $(n, C, gj (m), \xi)$ within Algorithm 4 will also satisfy that

$$\limsup_{n \to \infty} \frac{R(n; \mu, U, \rho EST}{\aleph^*(n; U)} < \infty. \tag{5.31}$$

Therefore, $O(\xi * (n; U))$ is the regret for the prospected ρEST strategy within absolute decentralized scenario with less information about the client number in an event when $\xi * (n; U) = \omega(logn)$. Subsequently, $O(f(n)logn)$ regret value is feasible for every function given that $f(n) \to \infty$ as $n \to \infty$. The subject pertaining the reason why the logarithmic regret can be conceivable within obscure client number might be intriguing. You should note down a clear distinction of the ρEST strategy according to Algorithm 4 within unidentified client number using the ρRAND strategy comprising identified client number as shown in Algorithm 3. Any function $f(n) = \omega(1)$, and the regret within the ρEST will be $O(f(n)logn)$, although the $O(\log n)$ is below the ρRAND strategy. Consequently, we are in a position of evaluating the deprivation of the execution processes during an event when the client number is not yet determined.

5.8 Impact of the Client Number and the Lower Bound Levels

5.8.1 Distributed Lower Bound Learning and Access

Presently, a well-supplied learning and accessible strategy has been structured comprising of evidence-based regret bound levels. Therefore, the relative execution process of the strategy is determined, contrasting them to the optimum learning and other access strategies. Therefore, all this is attained through analyzing the lower bound regret values of a particular uniform good policy based on a uniform class division rules. Thus the outcomes are restated below. Lower bound (Theory 6): Therefore, for a perfectly dispersed learning and access policy ρ, then the total regret sum in Eq. (5.2) thus satisfies Eq. (5.32) shown below

$$\lim_{n \to \infty} \frac{R(n; \mu, U, p)}{\log n} \geq \sum_{i \in U-\text{worst}} \sum_{j=1}^{U} \frac{\Delta(U^*, i)}{D(\mu i, \mu j^*)}. \qquad (5.32)$$

5.8.2 Client Number Behavior

Currently, the paper has already evaluated the total sum of the regret value of all strategies within the fixed number of clients U. Therefore it is high time we analyze the performance of the regret intensification as the U value rises on the other hand maintaining the channel number $C > U$ constant. Ranging client number (Theory 7): In an event when the channel number C is constant while the client number $U < C$ varies, then the total summation of the regret value based on centralized learning and access ρCENT will diminish while U rises and the higher limits within the sum regret within the ρRAND rise with U monotonically.

Therefore, in order to confirm that the sum of the regret value within the centralized learning and access diminishes with the client number U, it thus confirms to show for $i \in U -$ worst channel system, diminishes as U rises

$$\frac{\Delta(U^*, i)}{D(\mu i, \mu U^*)} \qquad (5.33)$$

You should note that $D(\mu i, \mu U*)$ and $\mu(U*)$ decline as U rises. Consequently, it also confirms showing that

$$\frac{\mu(U^*)}{D(\mu i, \mu U^*)}. \qquad (5.34)$$

declines while U increases. It is valid because it can be derived considering when U is negative. Considering the higher bound levels within the regret under the ρRAND if U is increased, then the U-worst channel number will decline and thus the initial term will diminish. The second phrase comprising collision $M(n)$ rises greatly.

5.8.3   Numerical Outcome

In this section, there is an analysis of simulations that vary in terms of schemes, client number, and the verification of the execution process of algorithms as exhibited in the sections pointed out earlier. The $C = 9$ channel systems are considered with the likelihood of the chances available depicted by Bernoulli supply comprising equally spaced bounds array from $0.1 - 0.9$.

5.8.3.1   Contrast of Dissimilar Schemes

Figure 5.2a rationalizes the regret value within the random and central allotted systems within the scene when $U = 4$ while every cognitive client contending for right of entry within the $C = 9$ channel systems. On the other hand, the lower bound theory for the regret value of the centralized case depicted from theory 2 together with the spread case depicted by theory 6 has also been plotted. Therefore the upper limit based on the random allotment case of theory 4 has not been plotted, because the limits are not tight since the client number rises every time.

Figuring out the taut higher bound is categorized as prospect work. Predictably it can be noted that the central allotment comprises the smaller regret value. Additionally, a very useful examination noted is the space amid the lower bound within the regret and the real regret which is found both in strewn and regional scenes. Within central cases, it is because of utilizing the g^{MEAN} data rather than using the optimum g^{OPT} data. Although within the spread scenarios, additional breaking point can be identified due to the fact that there is no account of any collision between each client. Therefore, the systems to be considered will be $O(\log n)$ and attaining the order optimum conditions despite the fact that they do not show optimum condition in the scaling constant.

Figure 5.2b clearly shows the logarithmic conduct of the increasing collision number. Although, the curve as shown in Fig. 5.2c considering the unidentified parameters tends to be linear in n because of the minimal n values. The disparity between g^{MEAN} and g^{OPT}: due to the disparity of the g^{MEAN} utilized in the paper based on the optimum data of g^{OPT}, there is execution of a simulation for clear comparison of the schemes within the data presented. Thus as prospected in Fig. 5.2b, there is high performance rates regarding the optimum scheme. Although, as pointed out in the earlier cases, the gMEAN permits individuals in availing a finite duration bound.

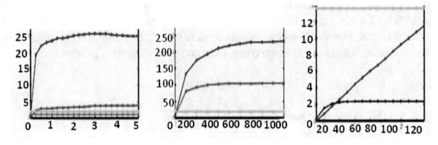

Fig. 5.2 (**a**, **b**, **c**): Simulation results. Probabilities of Available $\mu = \lfloor 0.1, 0.2, \ldots \ldots .0.9 \rfloor$

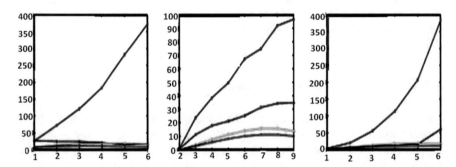

Fig. 5.3 (**a**, **b**, **c**): Simulation results. Probabilities of Available $\mu = \lfloor 0.1, 0.2, \ldots \ldots .0.9 \rfloor$

5.8.3.2   Execution Using Changeable U and C Values

The effect of escalating the number of secondary clients U based in the regret encountered by dissimilar rules through repairing the number of channels in C is shown in Fig. 5.3a. Therefore, with an increase in value of U, then there will be diminishing of the centralized scheme regret values and also increases for the supplied schemes as focused in theory 7. An increase in the number of collisions U will result in an increase in the monotonic regret value within the random allotment ρRAND [20]. On the other hand, an increase in the number of clients and a decrease in the U-worst channels will lead to a diminishing behavior of the monotonic value of the centralized case and also leading to a less regret value. Additionally, the lower bound levels for the distributed case at first will increase and then decline with U. The main reason behind this is because, since the U increases there will be two opposing results which include: an increase in the regret value as a result of an increase in the client number attending the U-worst channel systems and a decrease of the regret value as a result of a decrease in the number of U-worst channels.

Figure 5.3b analyzes the progress of dissimilar algorithms with the number of channels C varying and on the other hand correcting the client number U. Therefore, the likelihood of channel availability has been identified to be greater compared to

Fig. 5.4 Shows simulation results (Available probability $\mu = 0.1, 0.2 \ldots$, 0.9; the slots where users have excellent channels $U = 4, C = 9$, $n = 2500$ slots, 1000 runs, and ρRAND

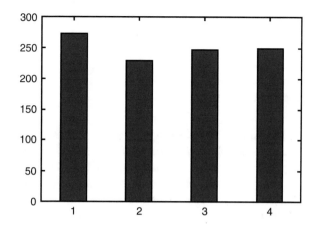

the one presented initially. In this case, the regret value increases monotonically with C in all the scenarios. An increase in the number of channels alongside the channel quality, the regret value rises due to a rise in the U-worst channels, and an increasing break point amid U-worst and the U-best channel systems. In addition, in a scenario where the ratio of UC is said to be fixed (0.5), while the channels and the client number and their quality will thus increase as depicted in Fig. 5.3c. An increase in the number of clients will lead to an increase in the regret value and the channel C number while its quality will also increase abruptly. Yet again, this is in compliance with the theory since the U-worst channel number rises as C and U increase, maintaining U and C in a fixed position.

5.8.3.3   Learning and Collision

According to Fig. 5.4, it clearly shows the nature of logarithmic collisions based on the unsystematic allotment scheme ρRAND. Moreover, the collision number is structured based on ρRAND centered on a perfect scene during an event when the channel data ease of access μ has been identified in viewing the impacts of learning on the number of collisions. Therefore, with a low value of the collision number extracted based on the identified channel parameters within the simulations is in compliance with the predicted theory evaluated as shown UE[$Y(U, U)$] in Lemma 2. Therefore, an increase in the number of slots as indicated by [21] will lead to an increase in the gap amid the collision number of the identified and the unidentified parameter because the initial would have converged to a finite constant as the latter increases as shown below.

Essential elements within the ρRAND are that it maintains fairness within the whole system. Consequently, it provides every client a chance of inhabiting in any place within the U-best channel systems. On the other hand, ρRAND's equal features are analyzed in Fig. 5.4. Additionally, the simulation presumes that $U = 4$ with cognitive clients campaigning of entry within the $C = 9$ systems. Also, the graph points out the clients who asymptotically acquire an excellent channel based on 1000 runs for the randomized allotted systems.

5.9   Summary

In the chapter, a novel strategy is presented based on the distribution of the learning channel available statistic and channel entry of several secondary clients within the cognitive networking system. Therefore, the initial strategy is presumed that the secondary client number within the networking system has been identified, although the second policy has eliminated the necessity. Additionally, it has also provided verifiable assurance for the strategies based on the total number of regret. Summing up the inferior bound regret of a particular perfect learning and accessible strategy, the initial strategy attains perfect optimum regret levels with the second strategy almost within the order optimum levels.

Therefore, the main investigation in this chapter is bringing out a clear reflection through the incorporation of various distributed and learning channels access systems within the cognitive practical networking system. Thus the main outcome in this paper provides fascinating collection of topics for future analysis. Consequently, there is need for the relaxation of the i.i.d. framework for the primary client transmissions and accurate detection of secondary clients which is also considered as an assumption. In this paper, the main strategy utilized permits unidentified although fixed secondary client number, since it is interesting integrating clients dynamically flowing in and getting out of the framework. Furthermore, the framework also pays no attention to the self-motivated traffic within the secondary client nodes together with the extension to the queuing theory is also admired.

References

1. Pimple, O., Saravane, U., Gavankar, N.: Cognitive learning using distributed artificial intelligence. Int. J. Mach. Learn. Comput. **5**(1), 7–11 (2015)
2. Anandakumar, H., Arulmurugan, R., Onn, C.C.: Computational intelligence and sustainable systems. In: EAI/Springer Innovations in Communication and Computing (2019)
3. Suganya, M., Anandakumar, H.: Handover based spectrum allocation in cognitive radio networks. In: 2013 International Conference on Green Computing, Communication and Conservation of Energy (ICGCE), Chennai, pp. 215–219 (2013)
4. Anandkumar, A., Michael, N., Tang, A., Swami, A.: Distributed algorithms for learning and cognitive medium access with logarithmic regret. IEEE J. Select. Areas Commun. **29**(4), 731–745 (2011)
5. Anandakumar, H., Umamaheswari, K.: Energy efficient network selection using 802.16g based GSM technology. J. Comput. Sci. **10**(5), 745–754 (2014)
6. Bonawitz, E., Denison, S., Griffiths, T., Gopnik, A.: Probabilistic models, learning algorithms, and response variability: sampling in cognitive development. Trends Cogn. Sci. **18**(10), 497–500 (2014)

7. Lan, Y., Cui, Z.: ILC with initial state learning for fractional order linear distributed parameter systems. Algorithms. **11**(6), 85 (2018)
8. Czarnowski, I.: Prototype selection algorithms for distributed learning. Pattern Recogn. **43**(6), 2292–2300 (2010)
9. Guo, Z., Lin, S., Zhou, D.: Learning theory of distributed spectral algorithms. Inverse Probl. **33** (7), 074009 (2017)
10. Heersmink, R., Knight, S.: Distributed learning: educating and assessing extended cognitive systems. Philos. Psychol. **31**(6), 969–990 (2018)
11. Pickett, M., Aha, D.: Using cortically-inspired algorithms for analogical learning and reasoning. Biol. Inspired Cognit. Architect. **6**, 76–86 (2013)
12. Nedjah, N., Macedo Mourelle, L.: Distributed learning algorithms for swarm robotics. Neurocomputing. 290–291 (2016)
13. Perlovsky, L., Kuvich, G.: Machine learning and cognitive algorithms for engineering applications. Int. J. Cognit. Inform. Nat. Intell. **7**(4), 64–82 (2013)
14. Al-Harthi, Y., Borst, S., Whiting, P.: Distributed adaptive algorithms for optimal opportunistic medium access. Mob. Netw. Appl. **16**(2), 217–230 (2010)
15. Anandakumar, H., Umamaheswari, K.: A bio-inspired swarm intelligence technique for social aware cognitive radio handovers. Comput. Electr. Eng. **71**, 925–937 (2018)
16. Anandakumar, H., Umamaheswari, K.: Supervised machine learning techniques in cognitive radio networks during cooperative spectrum handovers. Clust. Comput. **20**(2), 1505–1515 (2017)
17. Anandakumar, H., Umamaheswari, K.: Cooperative spectrum handovers in cognitive radio networks. In: EAI/Springer Innovations in Communication and Computing, pp. 47–63 (2018)
18. Haldorai, A., Ramu, A.: Cognitive social mining applications in data analytics and forensics. In: Advances in Social Networking and Online Communities (2019)
19. Wu, C.H.J., Tsai, J.H.: Concurrent asynchronous learning algorithms for massively parallel recurrent neural networks. J. Parall. Distribut. Comput. **14**(3), 345–353 (1992)
20. Kochen, M.: Representations and algorithms for cognitive learning. Artif. Intell. **5**(3), 199–216 (1974)
21. Gorsky, P.: Toward a unified theory of instruction in the cognitive domain. Int. Rev. Res. Open Distribut. Learn. (2007)

Chapter 6
Dynamic Spectrum Handovers in Cognitive Radio Networks

6.1 Introduction

Modern spectrum handover applications, as argued by [1], include radio spectrum distribution and sharing in a static capacity, in which the spectrum is designated to a specific essential (or authorized) client, for a significant timeframe with the end goal to evade obstructions. Parallel to this, to manage the escalating client requests, dynamic spectrum assignment for new remote systems is fundamental. Nevertheless, since existing remote systems involve broad parts of the radio spectrum, there is no adequate spectrum accessible to all the new unlicensed remote systems. Subsequently, a comprehensive analysis of these systems must be done to address this issue by means of dynamic sharing and task of spectrum. For instance, in the USA, Federal Communication Commission (FCC) considers to permit sharing of unused segments of TV groups to advance dynamic utilization of spectrum.

One compelling innovation to mitigate the issue of static spectrum task and to amplify dynamic spectrum use is Cognitive Radio (CR). This is a radio in present day remote frameworks, in which a CR (or an auxiliary client) hub changes its parameters (transmission or gathering) to share the spectrum dynamically and to keep away from the impedance with other essential or optional clients. The parameter change is finished by having some information about the radio condition factors, for example, radio recurrence (RR) signals, gadget level impedances (GLI), and so on [2]. To accomplish productive and dynamic allotment of spectrum between profoundly disseminated CR gadgets, a fair, straightforward, and helpful methodology is essential. Research is consequently on the advancement and investigating of the agreeable spectrum sharing methods in CR systems. Like CR organization, a multi-agent framework (MAF) is a framework made out of numerous independent operators, working separately or in gatherings (through association) to understand and comprehend specific undertakings. Like CR hubs, operators work dynamically to satisfy their client needs and no single specialist has a worldwide perspective of

© Springer Nature Switzerland AG 2019
A. Haldorai, U. Kandaswamy, *Intelligent Spectrum Handovers in Cognitive Radio Networks*, EAI/Springer Innovations in Communication and Computing, https://doi.org/10.1007/978-3-030-15416-5_6

the system. Every specialist keeps up its neighborhood view and offers its information (when required) with different operators to explain the doled out undertakings.

Late advances in innovation (particularly in the area of programmable coordinated circuits and disseminated man-made brainpower) have made an open door for us to build up another class of clever, self-ruling, and intuitive CR gadgets. These gadgets would then be able to be utilized in a wide assortment of system spaces (WLAN, WRAN, MANETs). Moreover, a proficiently planned CR with a product operator conveyed on it would be fit for associating with neighboring radios to frame a dynamic, inexactly coupled and infrastructure less community-oriented system. While CR physical engineering and its detecting capacities have gotten significant consideration, the topic of how to share radio assets in agreeable situations is likewise a vital research issue for spectrum flow specialists. In this way, in this paper, a MAS-based methodology is proposed for dynamic spectrum allotment. In particular, we think about agreeable MAS, in which the operators are sent over essential and secondary client gadgets. By agreeable MAS, we imply that the essential client specialists trade a tuple of messages and help neighboring auxiliary client operators to enhance their spectrum use.

Besides, the participation system we created is like the agreement net convention (ANC), whereby the individual auxiliary client (AU) operator ought to send messages to the suitable neighboring essential client (EC) specialists at whatever point required and, in this manner, the related (potential user) PU operators should answer to these operators with the end goal to make spectrum sharing assertions. We recommend that the secondary user (SU) specialists should take their choices dependent on the measure of spectrum, time, and cost proposed by the PU operators and should begin spectrum sharing at whatever point they locate a suitable offer (without holding up until the gathering of all the neighboring PU operators' reactions) [3]. At that point, after totally using the coveted spectrum, SU operators should pay the concurred cost to the regarded PU specialists.

Indubitably, this work is separated into following four sections:

- First, we present a concise best in class on different accessible methodologies for spectrum sharing utilizing multi-agent frameworks, amusement hypothetical methodologies, and medium access control arrangements.
- Second, we detail four distinct situations, in which spectrum sharing difficulties should be tended to in points of interest. We additionally propose some drive measures, which are important to be taken for productive usage of the accessible spectrum in the referenced situations.
- Third, we present a helpful structure with the related spectrum sharing calculations. The proposed MAS are helpful where PU specialists trade a progression of messages to impart their spectrum questioned by SU operators. The more perplexing situations with operators' focused practices will be inspected as a piece of our future investigation.
- Finally, we direct broad recreations to confirm the working of the proposed helpful calculations for dynamic spectrum partaking concerning psychological radio systems.

Whatever remains in the paper is composed as follows. The accompanying area quickly shows related works. A somewhat new use of multi-agent frameworks is for effective allotment of ghastly assets in CR systems. Fundamentally, the two know about their encompassing surroundings through cooperation, detecting, and checking and they have self-sufficiency and power over their activities and states. They can fathom the relegated errands autonomously dependent on their individual capacities or can work with their neighbors by having incessant data trades.

6.2   Literature Review

Research has been going around for quite a long while with the end goal to apply multi-agent frameworks for basic leadership process and asset sharing. A new utilization of multi-agent frameworks is for productive distribution of ghastly assets in CR systems. Essentially, the two know about their encompassing surroundings through communications, detecting, checking, and they have independence and power over their activities and states. They can explain the allocated undertakings autonomously dependent on their individual abilities or can work with their neighbors by having incessant data trades.

6.2.1   Detection of the Spectrum

An initial step is continuum detecting decides whether an essential client is available on a band. The spectrum, the subjective radio can impart the consequence of its discovery to other intellectual radios in the wake of detecting. The objective of spectrum detecting is to decide spectrum status and the authorized client's movement by occasionally detecting the objective recurrence band. Specifically, a subjective radio handset identifies a spectrum, which is unused, otherwise continuum gap, which include time, area, and band that identifies an accessing technique to (i.e., transmitting force and access span) without meddling of an authorized client's transmission. Spectrum detecting might be either unified or circulated.

In the incorporated spectrum detecting, a detecting director like the baseline stations and accessing points, which detect an objective recurrence orchestra, including the data got, remains imparted to different hubs in the framework. For instance, the detecting controller might be unable to identify the unconstrained client located at a cell edge. Either circulated continuum allocation, unrestricted clients sense the band freely, including the band detecting is accomplished utilized by individual psychological radios (non-agreeable detecting) or imparted to different clients (helpful detecting) [4]. Despite the fact that agreeable detecting manages a correspondence and handling overhead, the exactness of spectrum detecting is more noteworthy than that of non-helpful detecting. So, spectrum-detecting procedures might be arranged into three classes: Transmitter recognition, Cooperative location, and Interference-based identification.

6.2.2 Spectrum Management

With the end goal to meet the correspondence necessities of the clients' spectrum, the board catches the best accessible frequencies. The CRs to choose the best band of the spectrum with the end goal to meet the Quality of Service (QoS) wants all accessible recurrence groups; along these lines, the elements of the spectrum the board are critical for the CRs. These administration capacities can be delegated as follows.

6.2.3 Spectrum Investigation

The detecting of spectrum results is investigated to appraise the spectrum standard. Here one issue is the means by which to gauge the nature of spectrum accessed by a SU. Spectrum decision can portray this quality. Spectrum access requires a choice model. The unpredictability of this model is relied upon the parameters considered in the spectrum examination. The choice model turns out to be more mind-boggling when a Secondary User has numerous targets. For instance, a SU might need to amplify execution while limiting unsettling influence caused to the PU. A stochastic advancement technique is a fascinating device to show and take care of the issue of spectrum access in a Cognitive Radio [5]. Whenever clients (both essential and optional) are in the framework, inclination will influence the choice of the spectrum access. These clients can be agreeable or non-helpful in spectrum access. Every client has its very own motivation in a no cooperative domain. In a helpful one, all clients can cooperate to accomplish the objective. In an agreeable situation, CRs conform to one another.

6.2.4 Spectrum Mobility

The mobility of the spectrum is the capability that is denoted with multiple recurring spectrums of cognitive radio users. The moment the authorized user begins accessing the wireless systems that is in use with a restricted user, the unrestricted customer is capable of changing the inert spectrums into any functional band of spectrum. The transformation in working recurrence band may be referred to as the Handoff of Spectrum [6]. A measure in the diverse layer at the convection stack must be acclimated into coordinate the new working recurrence band amid spectrum handoff. Spectrum handoff must endeavor to guarantee that the unlicensed client can proceed with the information transmission in the new spectrum band.

6.2.5   *Spectrum Sharing*

Since there are a number of auxiliary clients who need to utilize accessible spectrum openings, an intellectual radio network needs to keep up the cooperation between its self-objectivity of data exchanging proficiently and caring objectivity to impart the accessible spectrum to other psychological and non-subjective clients. This is finished by approach rules which decide the conduct of intellectual radio networks in a wireless condition. The reasonable spectrum planning technique, open spectrum use in the spectrum sharing is one of the real difficulties.

6.2.6   *Dynamic Spectrum Access*

The access and ideology of dynamic spectrum identifies its identity and opening. Dynamic spectrum access is the most indispensable utilization of subjective radios. The PU groups are deftly accessed by the SU systems with the end goal that the obstruction caused to the PUs is immaterial. Figure 6.1 demonstrates the situation for dynamic spectrum access (DSA) where different PUs and SUs are operative together.

The approach is applicable with the wireless framework that adjusts into an accessible gap of dynamic spectrum licensed utility privileges. This is due to the transforming goals and status: made impedance that transits the status of radio into natural forms of limitations. DSA's principle assignment is the conquering of obstruction sorts, unsafe impedance due to gadget failing, including the hurtful obstruction due to malignant clients. Hence, three primary capacities are evident of the handover of dynamic spectrum. They are Cooperation of the spectrum, Subjective preparing,

Fig. 6.1 Concurrence of
numerous essential and
auxiliary client systems

Primary User Network

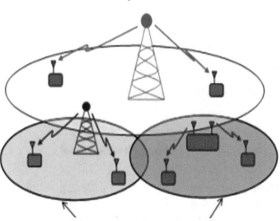

Secondary User Networks

and Access of the spectrum. Continuum cooperation stabilizes frequency of radio spectrum condition while handover of dynamic spectrum gives approaches headed for utilizing accessible continuum open doors aimed at reprocessing productively. Subjective handling is an insight in addition to basic leadership work, which plays out a few subtasks like learning of the radio condition, planning detecting productive, and access strategies, which oversees impedance for concurrence of the SU systems using the PU schemes.

6.2.7 Diverse Dynamic Spectrum Schemes and Models

The advanced assignment of the radio assets and small sharing of radio spectrum, which causes in spectrum deficiencies, the present spectrum the executives approach is made. In contrast with an immobile access of dynamic spectrum, the Dynamic Spectrum Access (DSA) is broadly utilized in the psychological system with different applications and methodologies.

6.3 Dynamic Exclusive Access

6.3.1 Utilize Display

The fundamental structure of the present spectrum control arrangement is kept up in this model: Spectrum groups are authorized for select utilize. The principal idea is to enhance spectrum productivity by presenting adaptability. Two methodologies are considerable in the model encompassing the rights and privileges of the spectrum and the handover of dynamic spectrum. The fundamental rights of the spectrum permit licensed users to exchange and offer spectrum, which necessitates selecting initiatives unreservedly. In this manner, markets and economies conform to an assumption of fundamental duties that apply assets that are more constrained. The approach of dynamic spectrum portions intends to facilitate a spectrum proficiency via dynamic spectrum tasks and application of transient and spatial movement insights by various administrations, i.e., spectrum is distributed to administrations for selective use per district in a certain timeframe.

6.3.2 Open-Sharing Prototypical

Open-sharing model is likewise called spectrum house demonstrate. In spectrum house show, each client has square with rights to utilize the spectrum. This, otherwise called open spectrum display, has been effectively connected for remote administrations, which works in the unlicensed mechanical logical and medicinal

(ISM) radio band (e.g., WLAN). Open sharing among clients as the establishment for dealing with a phantom locale is utilized by this model. There are three kinds of spectrum center models: (1) Uncontrolled-house, (2) Managed-lodge, and (3) Private-hall. (1) Uncontrolled-lodge: When a spectrum band is over crowding and utilizations the uncontrolled-house show, no substance has restrictive permit to the spectrum band. (2) Managed-house: Managed-center speaks to a push to maintain a strategic distance from the awfulness of hall by forcing a restricted type of spectrum handover structure. This asset is possessed and guided by a gathering of people or elements, and it is portrayed by limitations on when and how the asset is utilized.

6.3.3 Various Leveled Access Model

In various leveled access demonstrate, SUs utilize the essential assets with the end goal that the impedance to the private user is restricted. There are three methodologies in the model: Interweave, Underlay and Overlay. The weave show the possibility of an artful re-utilize the spectrum in the spatial space, i.e., the essential spectrum is used by CRs in the land zones where essential action is missing. Misuse of the supposed "spatial spectrum openings" is drawing in an enthusiasm, since numerous current authorized frameworks like, e.g., TV broadcasting and cell frameworks. Figure 6.2 demonstrates where "CR 1" can serve a portion of the SUs because no PU action is available in its vicinity.

Underlay: Underlay advances work in the utilized spectrum at a low power level for other authorized or permit excluded utilizes yet does not disable the clients. Underlay utilize is not authorized. Underlay access ideated CRs to work beneath the clamor floor of the PUs, including a propensity of Cognitive Radio correspondences without PUs monitoring.

Overlay: An overlay approach allows higher forces that could result in impedance to existing clients however defeats this plausibility by just allowing transmissions now and again or territories where the spectrum is right now unused.

Auxiliary CR client can transmit with a high transmitting capacity to expand their rates for giving spectrum openings in spectrum overlay approach, be that as it may, they need to locate the inactive recurrence groups which are unused by PUs. Likewise, in spectrum underlay approach, the SUs don't have to discover the spectrum openings and can transmit in the meantime coinciding with essential clients anyway they are not allowed to transmit with high transmitting force regardless of whether the whole RF band is inactive (whole RF is not utilized by essential clients) [7]. In this manner, Table 6.1 overlay is known as an impedance show where underlay is known as obstruction shirking model.

In writing, few strands of work have concentrated on spectrum sharing utilizing MAS. However, in these works, a few restrictions exist. For instance, MAS is utilized for data sharing and spectrum assignments. All the partaking specialists sent over access focuses (APs) frame a collaborating MAS, which is in charge of overseeing radio assets crosswise over gathered WLANs. The creators have not

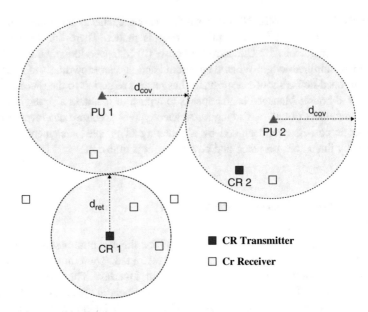

Fig. 6.2 Epitome of spectrum openings

Table 6.1 Comparison between an Agent and Cognitive Radio

Agent	Cognitive radio
Environmental awareness via past observation	Sensing empty spectrum portions and primary user signals
Acting through actuators	Deciding the bands and channels to be selected
Interaction via cooperation	Interaction via beaconing
Autonomy	Autonomy
Working together to achieve shared goals	Working together for efficient spectrum sharing
Contain a knowledge base with local and neighboring agent's information	Maintains certain models of neighboring primary users' spectrum usage

given any of the calculations and results for their methodology. The work considers an appropriated and dynamic MAS-based charging, evaluating, and asset allotment component where the specialists fill in as the salespeople and the bidders to share the spectrum dynamically. The convention utilized for radio asset designation between the CR gadgets and administrators is named as a multi-unit fixed offer closeout, which depends on the idea of offering and doling out assets. A definitive point of utilizing barters is to give a motivator to CR clients to expand their spectrum use (and henceforth the utility), while enabling system to accomplish Nash Equilibrium (an answer idea, where every client is accepted to know the balance procedures of alternate clients, and no client has anything to pick up by changing its very own methodology). Closeouts have conventional disadvantages of clients' untruthful practices, which can make genuine downsides the working of steadfast clients.

Diversion hypothesis has additionally been misused for spectrum portions in CR systems. In diversion hypothetical methodologies, each SU has one individual objective, i.e., to amplify its spectrum use and the Nash harmony is viewed as the ideal answer for the entire system (or amusement). Besides, it consolidates two fundamental suspicions: first, the sanity presumption, that is, the taking part essential and optional clients are objective so they generally pick procedures that amplify their individual gain. Furthermore, second, the clients' normal learning supposition, which incorporates the meanings of their inclination relationship. These presumptions may carry on well by permitting every client (or player) to judiciously settle on its best activity, in spite of the fact that in the majority of the aggressive recreations, here and there clients can give false data with the end goal to amplify their benefits and in this way can influence the entire system execution.

As indicated by some ebb and flow explore works, spectrum sharing issues are like medium access control (MAC) issues, where a few clients attempt to access a similar channel and their access ought to be imparted to the neighboring clients to keep away from the impedances. For the most part, in MAC-based spectrum sharing arrangements, when a CR client utilizes a channel, it sends a bustling sign to the neighboring clients through a control divert with the end goal to keep away from the obstruction. To evaluate control flags, the creator proposes a quick Fourier change-based radio plan, which empowers CR clients to identify the bearer recurrence of a control motion without making any destructive crashes the neighboring clients. Others recommend the utilization of a worldwide arrangement to trade the control data between CR gadgets [8]. Nonetheless, keeping up worldwide plans needs a lot of continuous data to be traded between CR clients causing complex gadget level structural overheads.

6.4   Spectrum Handover Scenarios

Here, we give a portion of the conceivable situations, which require the improvement of new answers for dynamic spectrum sharing. These situations are tended to as a piece of a Franco-German undertaking TEROPP. This task goes for creating different productive spectrum the board arrangements. Up to this point, our commitment to this venture is the improvement of an agreeable methodology for shrewd spectrum allotment. In these situations, the present spectrum assignments are static and inter-device impact is a major issue. Along these lines, effective arrangements are required with the end goal to empower dynamic spectrum utilization and to evade impedances. The situations are separated into four distinct areas as follows: (1) Spectrum sharing and impedance evasion in ISM groups, (2) Spectrum partaking in cell systems, (3) Opportunistic spectrum usage in TV groups, and (4) Spectrum assignment in specially appointed systems. In the wake of specifying and proposing conceivable activities towards dynamic spectrum access, we will portray our helpful system as an answer for empower spectrum sharing under impromptu system space. Exactly, multi-jump designs, topology changes, and landing and flight of hubs whenever are the explanations behind building up an agreeable answer for dynamic spectrum sharing under impromptu system setting.

6.4.1 Spectrum Sharing and Interference

Web home clients and organizations have received avoidance in ISM Bands. Recently, WLAN as a typical innovation. Portrayed by modest gadgets and sensible information rates, WLANs can be sent anywhere. Intended to work over permit free ISM (Industrial, Scientific, and Medical) groups, WLANs are confined to utilize just a couple of symmetrical channels, which is all that anyone could need to give remote access in a local location. Nonetheless, the enormous addition in the quantity of WLANs working in a similar area presented another obstruction level that could be anarchic. This obstruction is viewed as the fundamental impediment for WLANs execution and it acquaints new difficulties with all the neighboring innovations that work in the ISM groups. Comparative issues may emerge with the arrangement of LTE femto-cells [9]. These little cells, situated at a home or a building, can give better inclusion and higher limit in indoor conditions. In any case, they experience the ill effects of obstruction caused by the neighboring femto-cells. The basic purpose of presenting these two cases are the majority of the occasions undesirable and it should be maintained a strategic distance from.

As an obstruction evasion arrangement, we predict an agreeable situation where the gadgets in a WLAN or LTE cell can have CR capacities, which enable them to improve recurrence reuse. They can likewise choose an elective spectrum partition, if there should arise an occurrence of any obstruction [10]. At that point, they can send the recently looked spectrum parcel data to the neighboring gadgets with the end goal to evade the conceivable crashes.

6.4.2 The Sharing of Spectrum Within Cellular Networking Sites

Actually, the situation substantiates concepts relating to spectrum sharing with the aid of the cellular networking sites as administrated by a focal point (for instance, the base center) which is in a position of enacting fundamental issues to every client. Therefore, clients possessing CR abilities is purposed to execute signal estimations, which incorporate several behaviors with an end goal of effective utilization of the present spectrum range. Thus, such behaviors can be depicted as being behavioral policies, e.g., the use of the right MAC address, changing to a more suitable base center and the transmission of estimated reports to the center. Therefore, from the unique circumstance, circulated operational modes are considered special while the diverse overlay capacities might be executed, for example, the use of meeting offices, with the end goal of upgrading the frequency rate reuse and to empower proficient use of the accessible band ranges. Therefore, the best application scenario considered in this case is a hospital, since the population of the clients cannot be estimated precisely. Hence, considering the CR ability, a specific station can be in a position of identifying the ideal band range. Thus, the particular set of band can be directly linked and maintained using other adjacent gadgets by considering a series of progression with the aid of multi-agent frameworks and considering the quantity of the present clients and needs. Consequently, the mutual data band shared is conveyed to the BS specialists, which is used for managerial objectives.

6.4.3   TV Bands that Use Opportunistic Spectrums

Most European nations have been dealing with enhancing television facilities through halting of PAL communication signals and consequently using the DVB-T standards rather. Therefore, such a procedure can make an adequate measure of the under-utilized assets particularly because of the advanced profits. It gives us a chance to clarify the abuse of UHF bands in order to comprehend the numerical idea of all profits. For the most part, UHF groups are part of different channels, with channels 21–69 that were initially allocated to television administrations. Such direct channels are 8 MHz in width, and 21 channels compared to groups 470–478 MHz. The DVB-T is in a position of covering the entire neighboring divisions, and utilizations about 6 UHF groups through the communication of about 36 television channels. For instance, France comprises of almost 100 DVB transmitters, which are utilized during the broadcasting of these television channels. Within a specified range, it is postulated that the television channels to be used can be 6 amid 49 UHF centers, which excludes about 43 centers that have not been used. Therefore, such a tremendous measure of void ghastly assets legitimizes the global enthusiasm for most television channels.

During a gathering done under the global radio-communication meeting, various discourses regarding the usage of various television bands had already commenced. Most scholars chose to select UHF centers ranging from about 60–90 a global portable transmission media administrations. An alternative movement has been carried out by the European nations regarding the implementation of an assignment assemble TG4. Therefore, the TG4 is always in charge of estimating the execution of the DVB transmitters with the end goal of using the under-utilized bands within the television groups astutely. Additionally, the estimations will thus be contrasted while the outcomes achieved from the cell phones operating within the WiMAX bandwidth. In order to outline the results, the below section provides a couple of ventures that should be considered for an artful usage of the phantom assets within computerized profits as depicted below:

- Initially, the DVB transmitters should be in a position of attaining psychological radio detection abilities, observing, and portraying the underused television bands. Such conceivable bands' improvement comprises of productive signals' preparation procedures.
- At this point, based on the grounds that the DVB transmitters stake hugely their ghastly assets with the aid of the radio mouthpiece, in this way an exact range of sharing strategies should be conveyed.
- At last, few methods should be guaranteed with an end goal of separating the DVB-T and the amplifier's signal.

6.4.4 An Impromptu System with the Use of Spectrum Allotment

In this section, furnished SU, comprising of CR abilities and other special functionalities are utilized. It involves the client being exposed to more critical circumstances, which it does not have permission to the radio assets since its innovation processes are in need of a vitality, which most clients do not claim. The SU faculties with adjacent signs of essential clients PU1 and PU2 in stage one, coordinates with every operator sent as shown in stage two. Such participation procedures permit the SU to follow up on essential clients' reactions with accessible ranges in stage three. In this way, psychological radio capacities within the SU assume the character of interoperability with the end goal to acquire data pertaining to the neighboring clients' range groups and their entrance innovation [11, 12]. Moreover, these initiatives comprising of the specialists mainly coordinate and alter the SU's product design by stacking the essential calculations which best suited the present situation as shown in stage four in Figs. 6.3 and 6.4.

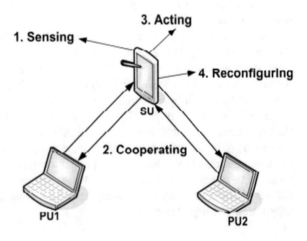

Fig. 6.3 An account of the Ad hoc system situation

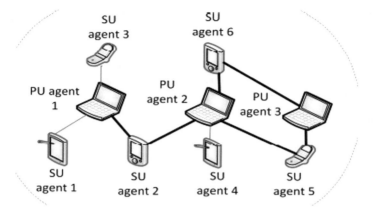

Fig. 6.4 An Ad hoc framework comprising of both the primary and the secondary clients

6.4.5 Main Issue Proclamation

According to the above situation, a job of an operator and a CR has been introduced in a specially appointed critical circumstance. In any case, thinking broadly and down to earth viewpoint, it is addressed that the range allotment of challenges within the private specially appointed zone or an all-around distinguished administrated edges, for example, grounds, gathering focus points, or a dispensary. One should note that such projected calculations likewise could be effortlessly connected to such crisis especially within the appointed system situation. According to our suggested situation in Fig. 6.3, an impromptu WLAN is shown which has been deployed in a set of essential PU = PU1, PU2 .... Pun with the subsequent SU = SU1, SU2.... Sun clients. In order to facilitate the conveyance of the hubs every operator should be sent in each set. In any point when the SU gadgets recognize the unfilled segments of the range required by the client, the specialists commence speaking to the PU operator till the point when the range sharing comprehension has been achieved.

6.4.6 Formalization

For example, If $G = N$, A, then the coordinated systems comprising of series of versatile hubs with an end goal $[(SU \, u \, PU)\epsilon N]$ comprises of arrangements of coordinated bends. For every coordinated curve as shown by points (I, j), there is an interface of $\in A$ of the optional client Sui which links it to the essential clients PU_j. Additionally, it can be symbolized by a coordinated circular segment $(j, i) \in A$ which demonstrates the bearing association between the PU_j and the Sui. Additionally, the optional client taking part with the neighboring clients comprise of a range of sharing arrangements. It is postulated that the measure of the optional client's ranges I wants to snatch from client j. Thus with t_{ji} as the measure of time i needs to use the range and P_{ij} as the value that will be paid for client j. On the other hand, primary client j, S_{ji} comprises of the measure of the range used to impart to I, while t_{ji} is thus the considered time limit and the P_{ji} value I is hoped to be achieved through the range sharing. Therefore, the above framework can be defined for each auxiliary client "i": as shown below:

Subject to Maximize

$$\sum_{(i,j)eA}^{\infty} (S_{ij}t_{ij}) \tag{6.1}$$

Subject to Minimize

$$\sum_{(i,j)eA}^{\infty} (P_{ij}) \forall SU \in N \tag{6.2}$$

Similarly for Primary users: Maximize

$$\sum_{(i,j)eA}^{\infty} (P_{ij}) \tag{6.3}$$

Subject to Minimize

$$\sum_{(i,j)eA}^{\infty} (S_{ij}t_{ij}) \forall PU \in N \tag{6.4}$$

And

$$l_{ij} \leq S_{ij} \leq U_{ij} \tag{6.5}$$

U_{ij} and l_{ij} comprises of the upper and the lower limits for the primary clients j. It means the secondary client not in the position of measuring the range that is higher than the stated point.

For instance, during static conditions, range divisions are allotted to essential clients and accordingly the web access suppliers will acquire the range costs. For instance, when considering the essential client, which is PU_j, he has purchased a vast segment of the range that is exactly 8 MB as shown in Fig. 6.3. Amid $t0$-$t1$, the doled out segment may stay occupied due to high number of client activity, for instance, the use of video conferencing and addressing, yet most of alternate occasions this range can remain not used ($t1$-$t2$ & $t3$-tn). The PU_j utilizes its range parcels for different exercises yet for most parts individuals they incline towards such sort of

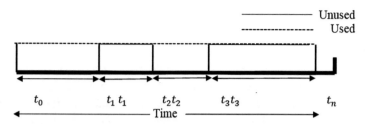

Fig. 6.5 Day V primary range usage

exercises that should be performed during week closes. Therefore, our projected elucidation provides the auxiliary client Sui who has the capability of picking an ultimate band channel progressively. Such a decision can be acquired in participation of a specialist set out using the PU_j, through the consideration of range measurement, with the regarded time range and other related cost put into consideration shown in Fig. 6.5.

6.5   Linked Resolutions Used for Active Spectrum Sharing

This segment mainly elaborates on the projected obliging range sharing arrangement, concisely, it comprises of both the secondary and the primary client's inner styles and various calculations of characters.

6.5.1   Section A: Agent

An agent is an inexactly coupled and dynamic section that comprises of abilities executing a certain task in a self-sufficient manner, in view of the information got from normal atmospheric conditions or potentially through other operations. Afterwards the approximated units cooperated in order to shape a multi-agent framework system. For the most part, specialists are more important and suitable for the SU hub, which is said that it allows the presentation of vast number of computerized reasoning of the CR systems and thus encourages the SU hubs in carrying out proficient activities through considering numerous links within the adjacent gadgets [13]. After they have been set up, it is agreeable that the multi-agent frameworks will be in a position of expanding the SU capacities in various assorted ways. Instantly, solitary SU operators can be constrained about learning data ranges relating to the spectrum knowledge. Yet heap of SU specialists who can wholly distinguish wide openings and deliver them in different hubs.

6.5.2   Section B: Contract

Few methodologies exist about collaboration of different net convention of various net multi-agent writings. Amid the select methodologies, CNP (contract net protocol) is the ultimate route to be used for operator collaboration and making wise decisions. The CNP explains the accumulation of different operators as a "contract net" with few specialists framing such nets with an end objective of elaborating doled out assignments. In this case, an operator can be a chief of a contractual worker. Fundamentally, and administrator can allocate an assignment to a temporary operator through sending a call of proposition (Cap's). Subsequently, other different qualified temporary workers demonstrate their enthusiasm through conveying their propositions in fathoming assignments. In this case, the supervisor will choose the finest proposition through acknowledgment and by granting the stated contract to the contractual worker. On the other hand, the temporary worker will explain the relegated undertaking as concurred by comparing with the administrator. Because of its basic and its productive nature, the stated methodology depends on reciprocal message trade and assignment allotment instruments of the use of CNP.

6.5.3   Section C: Operating Based on the Proposed Solution

Therefore, the SU stated structure comprises of the below accompanying five diverse interspersed components.

- The DSS (the dynamic range sensor) which is utilized in the detection of the unfilled range parts. Several techniques are available which is used for range detection, for example, the Pus feeble signal and its vitality identification, helpful incorporated locations, and so on [14]. It is important that the detection is executed relating to the DSS through thinking about a constant powerful condition, since it isn't clear the time a prospective band can get involved or can be released.
- The SC (the Spectrum characterizer), which highly portrays the range gaps dependent on Shannon's hypothesis for the creation of a limit-based plummeting requested rundown for every accessible PUs.
- The SUi is the third segment that conveys demand messages to the SU gadget specialist, at whatever point a client needs a bit of range.
- The AKM is the fourth part which gets the PU portrayal data from SC that acts as subsets of discharge having empty range spaces which are accessible. The rundown will be changeless but it will be refreshed and kept up on customary timeline interims dependent on the data given by the SC component. In addition to this, the AKM generates a CfP message dependent on the contributions of both the SC and the SUI. The SUID being the auxiliary client ID and thus used in aiding PU responses while comparing to the SUs which is a measure of the range usage, while d is the due date for acquiring the essential clients propositions.

- The ACM (i.e., Association for Computing Machinery) geo-throws the CFP to the adjacent PU operators. Accessibly discharged, it is implied that the PU specialists do not have a one-jump quarter and thus have unexploited the spectrum to stake. Additionally, ACM likewise is in charge of choosing the reasonable acquired proposition [15].

Being in a position of getting the CfPs, intrigued PU specialists will be in a position of conveying their recommendations relating to the SU operators. Therefore, such proposition will be in form of an accompanying structure: PUID, s, t, p (Proposal), where PUID is said to be the essential client specialist distinguishing proof. Moreover, s will be a measure of the PU range, which is provided regarding SU, t will be the postulated range holding duration, p will thus be PU value astute to be acquired. It should be noted that each PU specialist just comprises of both the ACM and the AKM segments, with AKM dealing with the adjacent area data while the ACM chooses the most reasonable Cfb by means of assisting each other. Every PU keeps up an arranged rundown of the CfP within its store dependent on the estimations of s and that of t for motivations behind future anticipations as shown in calculation two. Although in the meantime, local SU has brought up recommendations and acknowledged messages that are most reasonable to the proposition [16].

Therefore, the data connecting the chosen PU will be conveyed to the AKM for forthcoming associations. If there should arise an occurrence of an acknowledged message from the chosen SU, a range sharing will be carried out dependent on concurred parameters on both parties [17]. Although the PU presently can react to the CfPs on the off chance that it needs its other idle range bits to be pooled. On the off chance that the PU gets a negative message from the SU, then it will keep on conveying proposition to assist accessible CfPs that the due date did not lapse [18]. In the section stated above, there was an introduction of an agreeable system used for a range assignment that easily produces exceptionally powerful conduct in unique situations and accomplish enhanced effectiveness of taking part specialists. Projected arrangement will thus depend on multi-agent framework participation through sending of operators over essential and auxiliary clients. Various exploratory assessments exhibited in the accompanying segment affirm to the effectiveness of the stated dynamic range assignments [19].

6.6   Experimental Results

This segment presents the reproduction outcomes directed with the end goal of improving the working and execution of the proposed range assignment calculations. The papers commenced by inspecting the accomplished utility of the essential and auxiliary clients and afterward think about the time esteems of the range used. Words are utilized reciprocally all through the accompanying areas.

6.6.1   Section A: Reproduction Setup

A reenactment role is played under the supposition of a quiet and versatile specially appointed system. Portable impromptu implies that the hubs in the area of every one of the SUs change. Arbitrarily, we put various essential and optional clients in a predefined region where every gadget contains a specialist sent over it for collaboration purposes. Additionally, two distinctive settled estimations of times, (i.e., $T1$ and $T2$) are accepted, with Time one ($T1$) speaks to the transient case while Time two ($T2$) will have a more extended period. At the point when $T1$ is considered, SU specialists can request a measure of range within 1 h limit (e.g., 0 is less or equal to $T1$ is less or equal to 60 min). While corresponding limit inside 2 h, as if there should arise, an occurrence of $T2$ (e.g., 0 is less or equal to $T1$ is less or equal to 120 min).

Such estimations provide similar measure of time esteems in genuine remote conditions without digging into complex circumstances. Our reproduction begins with the aggregate number of six SUs and four Discharge, and for each next round, there is an expansion of ten operators. The reenactment is directed for ten resulting rounds, with an aggregate of 20 h out of every day, for both $T1$ and $T2$ separately and the normal estimations of parameters are taken to draw the diagrams. The PU specialist's utility is determined as the cost paid by SU operators for range use partitioned by the measure of range it has shared for the regarded era as required by the SUs. SU operator's efficacy is spoken to as its range utilization for the required time isolated by the relating value paid to the Discharge. Accordingly, by appointing weights or needs to every one of the referenced parameters, the proper utility qualities for both the essential and auxiliary clients are picked.

It is presumed that each subsequent PU has an irregular accessible range divides and the area of SUs and Discharge is arbitrarily evolving. Additionally, we pursue the supposition that once concurred, Discharge would not have the capacity to pull back their responsibilities, and they should impart their range to the comparing SUs for the concurred era. Further, the aggregate number of participation messages created in the framework decides the collaboration cost. Along these lines, the collaboration procedure that is better as far as less number of messages that gives great utility qualities is considered as the most cost proficient. The aggregate number of assets effectively shared presents the achievement rate, while the quantity of no allocated range partitions measures the general range misfortune Shown in Fig. 6.6.

6.6.2   Results and Performance

Therefore, Fig. 6.6 also shows an analysis of the normal utility of every essential and optional client at $T1$ with those at $T2$ for various quantities of clients from 10, 20, 30,.... and so on. Fig. 6.6 portrays that when time limit is $T2$, the utilities are somewhat less contrasted with the outcomes got at $T1$. This is because the earth is versatile and a portion of the clients is somewhat reluctant to share their range for longer periods. We see that when there are 10 specialists, the normal utility qualities

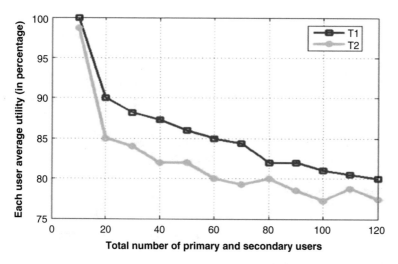

Fig. 6.6 Percentage utility of the operators

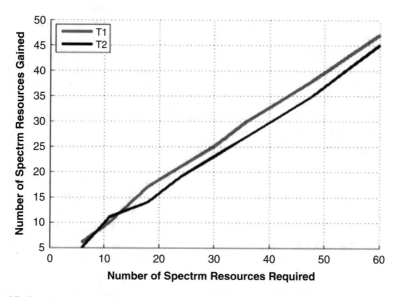

Fig. 6.7 Requirement and utility of the spectrum source and the SU

are relatively indistinguishable for both *T*1 and *T*2, demonstrating the ideal conduct. In any case, in different cases, the normal utility qualities are unique, demonstrating that the execution of specialists as far as their normal utility qualities has diminished marginally with the expanded number of operators.

On the other hand, Fig. 6.7 shows the range asset prerequisites and use after some *T*1 and *T*2 periods. Before all else, every one of them is totally shared; though when the required range assets land at the center qualities, around 90% of them are shared.

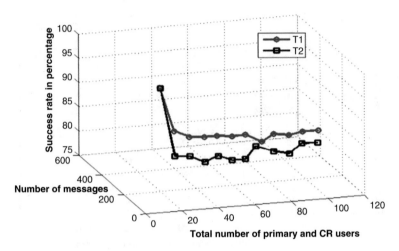

Fig. 6.8 Successful message rate

Such range sharing pattern keeps following a similar example achieving greater qualities, with accomplished total of assets involved somewhere in the range of 45 and 50. Therefore, the execution range sharing is high, even with expansive asset necessities. Our methodology is additionally in respect to time, in light of the fact that in CR arranges the range holding time is a standout among the most critical components to be considered. Once more, we run the reproduction with a few estimations of essential and auxiliary client operators. At the point when time limit is $T1$, the outcomes are completely fulfilled, for 80–120 specialists, while the time esteems are to some degree lesser at $T2$. Both the outcomes are super direct and intelligent with those of Fig. 6.7, which shows that the range sharing stays high even with the bigger number of operators.

According to [13], there is a delineation of the extreme number of bolstered SUs by the neighboring Discharge. Upheld SUs are those, which have totally picked up the required range. We see that when there are 10–15 Discharges, the quantity of upheld SUs is truly the equivalent for both $T1$ and $T2$. This implies, for set number of specialists regardless of whether the time esteems are high, the quantity of bolstered SUs is nearly the equivalent. Nevertheless, with huge number of operators, the quantity of upheld SUs at $T2$ is slimly lesser than $T1$. Hence, in impromptu circumstances, on the off chance that we increment the time esteems alongside an addition in number of operators, the outcomes will be somewhat less ideal. The quantity of participation messages transmitted and got in the whole framework with the achievement rate is appeared in Fig. 6.8 and Table 6.2. We can likewise observe that the methodology is direct as far as messages and achievement rate. Especially when time limit is $T2$, the execution of the methodology considerably debases, however it stays relentless.

A critical part of the methodology is the investigation of execution of changes as the measure of taking interest in specialist's increments. An increase of the operator

Table 6.2 Number of messages and success rate at T1 and T2

No. of agents	Number of messages		Success rate (in %)	
	T1	T2	T1	T2
10	45	41	100	98.7
20	81	72	90	85
30	117	115	88.23	84
40	159	161	87.31	82
50	185	176	86	82
60	253	261	85	80
70	271	262	84.41	79.3
80	325	366	82	80
90	388	392	82	78.53
100	416	434	81	77.26
110	475	483	80.5	78.77
120	503	516	80	77.42

demands, the level of range misfortune develops on a relentless pace. It is on the grounds that a portion of the SUs is not ready to discover non-occupied Discharge or because of the relative change in their locality. According to the illustrated figures in Table 6.2 shown, it is likewise certain that the measure of in general range misfortune is least 10–15%, if the quantities of clients are at the center phases. Range misfortune at this point achieves bit higher qualities, with increment number of operators, yet there is definitely not a quick debasement in the general framework execution. Alternate factors, for example, crashes, gadget level obstructions and postponements have not been mentioned here.

6.7 Discussion

These examinations and outcomes demonstrate viability of our answers with one end goal of providing dynamic spectrum handover to CR systems, which can endow better utility of specialists operating under minimal participation messages. Nonetheless, there are some imperative focuses identified with these outcomes, which require further research. To begin with, we accept that the impromptu condition is without obstruction; in any case, this presumption is not fully exploited. In actuality, the transmission intensity of gadgets is high to the point that they can undoubtedly intrude on the working stream of neighboring gadgets, causing impedances.

Subsequently, tending to spectrum handover under impedance empowered impromptu systems is a pending issue and a few specialists comprehend this issue to the considerable point of interest. Consequently, it is identified with the predetermined number of operators we have used to play out our analyses. Since, JADE enables a greatest number of 100–120 specialists on a solitary machine; in this manner, we have just demonstrated the conduct of our methodology with set numbers of operators. With the end goal to demonstrate the predictable working of our model with huge number of operators, we are dealing with creating scientific model dependent on the Markov chain. This model likewise assists correspondence rate, operators' value, and verification of different limitations.

6.8   Summary and Future Prospects

In conclusion, dynamic structure of spectrum handover can produce exceedingly viable conduct in various situations and accomplish better utility cognitive gadgets. The proposed methodology in this research depends on multi-agent systems participation executed by sending specialists on subjective radio-essential client gadgets. Test assessments affirm the productivity of the calculations for dispersed and decentralized situations. The results demonstrate that the proposed methodology can ingest a dynamic spectrum handover request by presenting the participation among essential and auxiliary client gadgets. Besides, the proposed methodology enhances the general utility and limits the spectrum adversity with a base correspondence cost.

Spectrum handover achievement rate is relatively 80% even with a substantial number of specialists [5]. We just proposed an explicit collaborative methodology to amplify the framework utility; the proposed participation structure can be stretched out towards limiting other key issues, for example, covering other clients' obstructions and impacts. This means that the issue is viewed as a piece of the proceeding works. Presently, we are dealing with scientific investigation of our methodology utilizing Markov chain. Additionally, the proposed methodology expects that network hubs are significant while in genuine frameworks, hubs can be insignificant, so further research is required to investigate the more spectrum handover practices.

References

1. Hu, H., Zhou, W., Song, J.: Dynamic spectrum sharing scheme based on spectrum adaptation and MIMO-OFDM in cognitive radio. J. Electr. Inform. Technol. **30**(7), 1548–1551 (2011)
2. Hakim, K., Jayaweera, S., El-howayek, G., Mosquera, C.: Efficient dynamic spectrum sharing in cognitive radio networks: centralized dynamic spectrum leasing (C-DSL). IEEE Trans. Wirel. Commun. **9**(9), 2956–2967 (2010)
3. Ge, Y., Sun, Y., Jiang, H., LI, J., LI, Z.: Research on dynamic spectrum allocation using cognitive radio technologies. Chin. J. Comput. **35**(3), 446–453 (2012)
4. Anandakumar, H., Umamaheswari, K.: Supervised machine learning techniques in cognitive radio networks during cooperative spectrum handovers. Clust. Comput. **20**(2), 1505–1515 (2017)
5. Haldorai, A., Ramu, A.: Cognitive social mining applications in data analytics and forensics. In: Advances in Social Networking and Online Communities. IGI Global, Hershey, PA (2019)
6. Wei, N., Chen, Z.: Cognitive radio as enabling technology for dynamic spectrum access. Appl. Mech. Mater. **347-350**(4), 1716–1719 (2013)
7. Song, M., Xin, C., Zhao, Y., Cheng, X.: Dynamic spectrum access: from cognitive radio to network radio. IEEE Wirel. Commun. **19**(1), 23–29 (2012)

8. Da, B., Ko, C.: Dynamic spectrum sharing in orthogonal frequency division multiple access—based cognitive radio. IET Commun. **4**(17), 2125 (2010)

9. Moon, B.: Dynamic spectrum access for internet of things service in cognitive radio-enabled LPWANs. Sensors. **17**(12), 2818 (2017)

10. Garhwal, A.: A survey on dynamic spectrum access techniques for cognitive radio. Int. J. Next Gener. Netw. **3**(4), 15–32 (2011)

11. Zhu, K., Wang, P., Han, Z.: Dynamic spectrum leasing and service selection in spectrum secondary market of cognitive radio networks. IEEE Trans. Wirel. Commun. **11**(3), 1136–1145 (2012)

12. Park, J., Chung, J.: Prioritized channel allocation-based dynamic spectrum access in cognitive radio sensor networks without spectrum handoff. EURASIP J. Wirel. Commun. Netw. **2016**(1), 50 (2016)

13. Gupta, V., Kumar, A.: Wavelet based dynamic spectrum sensing for cognitive radio under noisy environment. Proc. Eng. **38**(1), 3228–3234 (2012)

14. Lin, Y.-E., Liu, K.-H., Hsieh, H.-Y.: On using interference-aware spectrum sensing for dynamic spectrum access in cognitive radio networks. IEEE Trans. Mob. Comput. **12**(3), 461–474 (2013)

15. Zhu, X., Zhu, H.: Synergy routing and dynamic spectrum allocation in multi-hop cognitive radio networks. IET Netw. **3**(2), 82–87 (2014)

16. Vimal, S., Kalaivani, L., Kaliappan, M.: Collaborative approach on mitigating spectrum sensing data hijack attack and dynamic spectrum allocation based on CASG modeling in wireless cognitive radio networks. Clust. Comput. **4**(8), 100 (2017)

17. Anandakumar, H., Umamaheswari, K.: Cooperative Spectrum handovers in cognitive radio networks. In: EAI/Springer Innovations in Communication and Computing, pp. 47–63 (2018)

18. Klumperink, M., Shrestha, R., Mensink, E., Arkesteijn, V., Nauta, B.: Cognitive Radios for dynamic spectrum access—polyphase multipath radio circuits for dynamic spectrum access. IEEE Commun. Mag. **45**(5), 104–112 (2007)

19. Anandakumar, H., Umamaheswari, K.: A bio-inspired swarm intelligence technique for social aware cognitive radio handovers. Comput. Electr. Eng. **71**, 925–937 (2018)

Chapter 7
Supervised Machine Learning Techniques in Intelligent Network Handovers

7.1 Introduction

In computer science, machine learning is a critical and fast-growing field, which has an advance array of application. This field denotes automated exposure of fundamental sets and patterns of data. Tools used in machine learning are interlinked to the enhanced programs that have the capabilities of learning and adapting. Being the mainstay in the field of information technology, machine learning has become a central aspect of research considering a rapidly increasing datasets. The increased data implies that smart analysis of data will be pervasive and a fundamental ingredient in technological advancements. There are many applications that apply various techniques of SML, data mining being one of them. Users are fond of committing mistakes when analyzing or establishing the interlinking between different computing features. SML and Data Mining (DM) are denoted as Siamese partners, which provide an evidence of effective machine learning algorithms. The field of DM and machine learning has witnessed tremendous and drastic transition due to the aspect of evolution of nano and smart technology that have led to development of hidden data patterns to deriving different values.

Machine learning, information theories, statistical fusion, and computation have formulate solid technological and scientific assumptions that are traced from a firm calculation basis and sophisticated tools. Algorithms in machine learning are formulated into taxonomies considering the speculated algorithm results [2]. SML delivers the functions and frameworks which map an input to the correct output. The generation of unmatched data has formulated the SML techniques to be complicated progressively, which has initiated different algorithms for unsupervised and supervised machine learning. SML is relatively fundamental in categorized issues due to its obligation to instigate learning in computation and categorization system formulated. Machine learning is purposed for attaining an ease of access veiled in Big Data. The handover of SML on assured extraction of distinct and big data source

© Springer Nature Switzerland AG 2019
A. Haldorai, U. Kandaswamy, *Intelligent Spectrum Handovers in Cognitive Radio Networks*, EAI/Springer Innovations in Communication and Computing, https://doi.org/10.1007/978-3-030-15416-5_7

fundamentality is minimally dependent on the scheduled personal track and data-enhanced on machine scaling.

7.2 Supervised Machine Learning Background

SML algorithms is denoted though observable explanations and experiences, in consideration of performance measures and tasks classification. Methods used to evaluate machine learning as critical in molecular biological data due to its learning capabilities of algorithms to formulate hypotheses and classifiers, which valuates a complex relation with datasets. Hypotheses and classifiers are interpretable on a field specialist tasked with suggesting wet-lab experiments to refute or validate a hypothesis. The feedback loop, which is a critical aspect of SML in bio-informatics, interlinked between the vivo/vitro and silico experiments speeds up the process of knowledge recovery on biological data. The general tasks for learners are to classify, cluster, and characterize input data. For biological issues, categorization is a critical and common task, whereby $E+$ and $E-$ ($E + \cap E- = \varnothing$). In this scenario, the learner is expected to formulate a classifier, which distinguishes the negative set and positive examples. The classifier is utilizes as a categorization basis of unrevealed dataset for the future. Resultantly, for the case of a supervised categorization issue, the formulated and trained samples in a tuples set from $\{(,) \ldots (,)\}$ 1 1j n nj x y x y, whereas xi denotes the label of the class and y ij denotes the attribute sets of relevant instances. Considering the obligation of studying algorithms to deliver the classifier (function and hypothesis) is aimed at categorizing various instances into their relevant classes. However, this analysis only considers the SML applicable in categorization [3].

7.3 Methodologies and Literature

7.3.1 Supervised Machine Learning Algorithms

The article was written prior to the performance of an empirical evaluation of a principle-based learning framework (One rule, decision tree and the decision system). The analysis also included an analysis of statistical frameworks, namely Instance Based, Naïve Bayes, and the Neural Networks in consideration of the ensemble methodologies like the Bagging, Boosting, and Stacking of datasets based on the positive predictive values, accuracy, sensitivity, and specificity of machine learning algorithms. The learning methodologies applied in this analysis were retrieved from the WEKA learning package.

7.3.2 Supervised Datasets

Datasets from the UCI machine learning depository were utilized in this research considering the biological analysis and motivation of the relative datasets: *E. coli* datasets with an objective of predicting the localization of cellular sites of the *E. coli* proteins. These cellular sites are eight in number: cytoplasm (cp), periplasm (pp), the inner membrane lacking a signal sequence (im), the inner membrane having un-cleavable sequence of signals (imU), the outer membrane, outer lipoprotein membrane, inner lipoprotein membrane (imL), and the inner membrane composed of the cleavable sequence of signal (imS) [4]. The sequence of signal are denoted by the attributes, which are recognition methodologies like the Von Heijne and McGeoch with the availability of the N-terminal of the lipoprotein predicted, including the three various scoring functions on relevant content of amino acid. These sets are either predicted as an inner membrane, outer membrane, cleavable or the un-cleavable signals of sequence.

The Yeast Dataset—The yeast dataset is only fundamental in the articles due to its objectivity to determine the localization of the cellular membrane of the relevant yeast protein. Assuming similar objectivity as the E. coli dataset, the yeast dataset has ten various sites, namely: Nuclear, Cytosolic, Mitochondrial, Membrane Protein, Membrane Protein (Non-Terminal Signal), Membrane Protein (the Cleaved Signal), Extracellular, Vacuolar, Peroxisomal (POX), and the Endoplasmic Reticulum Lumen.

The Promoter Dataset—The errand of the classifier is to anticipate whether a DNA succession from E. coli is either an advertiser or not. The data input is a 57-nucleotide arrangement (A, C, T, or G).

HIV informational collection—The informational index contains 362 octamer protein arrangements every one of which should be delegated a HIV protease cleavable site or un-cleavable domain.

7.4 Fundamental of SML Models

Maybe the most effortless conceivable calculation is the straight regression. Once in a while this can be graphically denoted as a straight line, however regardless of its name, if there is a polynomial theory, this line could rather be curved. In any case, it shows the connections between scalar ward variable "yy" and at least one informative qualities indicated by xx. In layman's terms, this implies direct relapse is the calculation which takes in the reliance between each known xx and "yy", with the end goal that later we can utilize it to foresee yy for an obscure example of xx. In our first directed learning precedent, we will utilize a fundamental straight relapse model to foresee an individual's circulatory strain given their age [5]. This is an exceptionally straightforward dataset with two significant highlights: Age and circulatory strain. As of now referenced above, most machine learning calculations work by

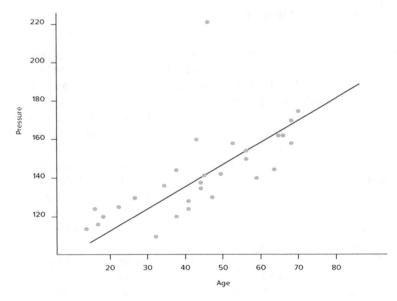

Fig. 7.1 Data sample and hypotheses

finding a factual reliance in the information gave to them. This reliance is known as a theory and is typically meant by h (θ) h (θ). To make sense of the theory, we should begin by stacking and investigating the information.

In the Fig. 7.1 shown above, each blue dab denotes the information test and the blue line is the speculation which our calculation needs to consider. So what precisely is this theory at any rate? To take care of this issue, we have to take in the reliance among xx and yy, which is indicated by y = f(x) y = f(x). Along these lines f(x) f(x) is the perfect target work. The machine learning calculation will endeavor to figure the theory work h(x) h(x) that is the nearest estimate of the obscure f(x) f(x). The least difficult conceivable type of speculation for the direct relapse issue resembles this: h θ (x) = θ0 + θ1∗ x h θ (x) = θ0 + θ1∗x. We have a solitary info scalar variable xx which yields a solitary scalar variable yy, where θ0θ0 and θ1θ1 are parameters which we have to learn. The way towards fitting this blue line in the information is called straight relapse. Understand that we have just a single info parameter x1x1; be that as it may, a great deal of theory capacities will likewise incorporate the predisposition unit (x0x0). So our subsequent theory has a type of h θ(x) = θ0∗x0 + θ1∗x1hθ(x) = θ0∗x0 + θ1∗x1. Be that as it may, we can abstain from composing x0x0 on the grounds that it's quite often equivalent to 1. The theory looks like h(x) = 84 + 1.24xh(x) = 84 + 1.24x, which implies that θ0 = 84θ0 = 84 and θ1 = 1.24θ1 = 1.24. By what means can we naturally determine those θ values? We have to characterize a cost work. Basically, what cost work does is to just ascertain the root mean square mistake between the model expectation and the genuine yield.

$$J(\theta) = \frac{1}{2m} \sum_{i=1}^{m} \left(h_\theta \left(x^{(i)} \right) - y^{(i)} \right)^2 \qquad (7.1)$$

For instance, our speculation predicts that for somebody who is 48 years of age, their circulatory strain ought to be h (48) = 84 + 1.24∗48 = 143 mmHg h (48) = 84 + 1.24∗48 = 143 mmHg; in any case, in our preparation test, we have the estimation of 130 mmHg. In this manner, the blunder is (143–130)2 = 169 (143–130)2 = 169. Presently, we have to compute this mistake for each and every section in our preparation dataset, at that point total it together $J(\theta) = \frac{1}{2} m \sum_{i=1}^{m} \left(h_\theta \left(x^{(i)} \right) - y^{(i)} \right)^2$ and remove the mean an incentive from that. This gives us a solitary scalar number which speaks to the expense of the capacity.

We will likely discover $\theta\theta$ qualities to such an extent that the cost work is the most reduced; in alternate words, we need to limit the cost work. This will ideally appear to be instinctive: If we have a little cost capacity esteem, this implies the blunder of expectation is little too. Presently, we have to discover such estimations of $\theta\theta$ to such an extent that our cost work esteem is negligible.

$$\min J(\theta) = \frac{1}{2m} \sum_{i=1}^{m} \left(h_\theta \left(x^{(i)} \right) - y^{(i)} \right)^2 \qquad (7.2)$$

There are a few conceivable calculations, yet the most prominent is inclination drop. So as to comprehend the instinct behind the slope drop technique, allows first plot it on the diagram. For effortlessness, we will accept a less complex speculation h (θ) = θ1∗ x h (θ) = θ1∗ x. Next, we will plot a straightforward 2D graph where xx is the estimation of $\theta\theta$ and yy is the cost work shown in Fig. 7.2.

The cost work is curved Fig. 7.3, which implies that on the interim [a,b][a,b] there is just a single least. This again implies that the best $\theta\theta$ parameters are at the point where the cost work is insignificant. Essentially, inclination plummet is a calculation that attempts to locate the arrangement of parameters which limit the capacity [6]. It begins with an underlying arrangement of parameters and alliteratively makes strides in the negative heading of the capacity inclination.

On the off chance that we ascertain the subsidiary of a theory work at an explicit point, this will give us a slant of the digression line to the bend by then. This implies that we can compute the incline at each and every point on the diagram. The manner in which the calculation works is as follows:

- We pick an arbitrary beginning stage (irregular $\theta\theta$)
- Calculate the subordinate of the cost work now
- Take the little advance towards the incline θj:= θj − λ∗ $\partial\partial\theta j$ ∗ J(θ)θj:= θj − λ ∗ $\partial\partial\theta j$ ∗ J(θ)
- Repeat stages 2–3 until the point that we join.

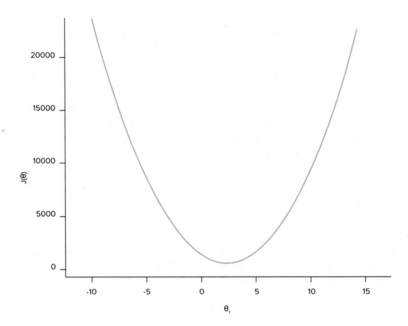

Fig. 7.2 Cost function

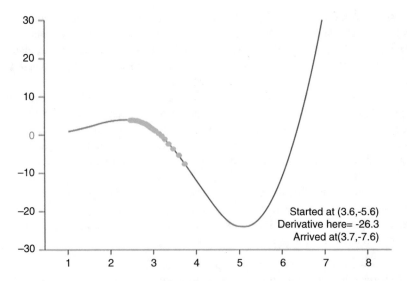

Fig. 7.3 Descending with step coefficient 0.005

7.5   Forms of Statistical Data

When working with information for machine learning issues, it is imperative to perceive diverse sorts of information. We may have numerical (ceaseless or discrete), clear cut, or ordinal information. Numerical information has importance as estimation. For instance, this includes age, weight, number of bitcoin that an individual possesses, or what number of articles the individual can compose every month. Numerical information can be additionally separated into discrete and persistent sorts. Discrete information can be checked with entire numbers, e.g., number of rooms in a flat or number of coin flips. Constant information can't really be spoken to with entire numbers. For instance, in case you're estimating the separation you can hop, it might be 2 m, or 1.5 m, or 1.652245 m [7]. Information show the qualities, for example, individual's sexual orientation, conjugal status, nation, and so forth. This information can take numerical esteem; however, those numbers have no scientific importance. You can't include them together. Ordinal information can be a blend of the other two sorts, in that classes might be numbered in a scientifically significant manner. A typical model is evaluations: Often we are requested to rate things on a size of one to ten, and just entire numbers are permitted. While we can utilize this numerically, e.g., to locate a normal rating for something we regularly treat the information as though it were absolute with regard to applying machine learning strategies to it.

Machine learning is fine, appropriate, towards the unpredictability of dealing with through divergent information beginning and the immense scope of factors and in addition measure of information concerned where ML flourishes on expanding datasets. The additional information given to the ML structure, the more it have the capacity to be prepared and concern the results to unrivaled estimation of bits of knowledge. At the freedom from the limits of individual dimension thought and study, ML is shrewd to discover and demonstrate the examples covered up in the information. One standard definition of the managed learning assignment is the arrangement issue: The student is required to learn (to surmise the conduct of) a capacity which maps a vector into one of a few classes by taking a gander at a few information/yield instances of the capacity. Inductive machine learning is the way towards taking in a lot of guidelines from occasions (models in a preparation set), or all the more as a rule, making a classifier that can be utilized to sum up from new occurrences. The way towards applying directed ML to a true issue is depicted in the Fig. 7.4 below.

7.6   Supervised Machine Learning Techniques

The fundamental obligation in machine learning technicians is to formulate the network intrusion determining framework [8]. Thus, this article considers the evaluation and performance of the relevant classifiers: Support Vector Machine, Naïve

Fig. 7.4 SML processes

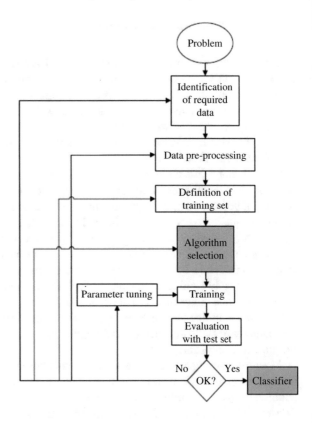

Bayes, Logistic Regression and the Random Forest in reference to intrusion uncovering. The relevant classifiers include:

7.6.1 Logistic Regression

The elemental purpose of this classifier is to object the categorization issues. Logistic regression operates for both the multi-class and binary class categorization, whereby the speculation of event occurrence determines through the fitting datasets on the logistic functioning. The resultant valuation denotes through the logistic functioning ranges from value zero to one. For instance, when the valuation is 0.5 or more the label would read zero Fig. 7.5.

The output equals to zero or one while the hypothesis Z equals $WX + B$. The Hypothesis x equals sigmoid Z. When "Z" is to the infinity, the predicted (Y) becomes 1 and when "Z" becomes the negative infinity, the predicted becomes zero.

Fig. 7.5 The sigmoid
function

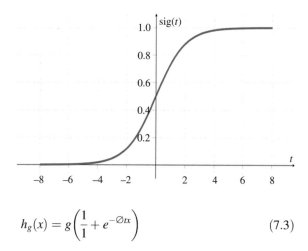

$$h_g(x) = g\left(\frac{1}{1} + e^{-\emptyset tx}\right) \tag{7.3}$$

7.6.2 Support Vector Machine

In most instances, the Support Vector Machine is used as a classification algorithm problem, but is also being applied as a regression problem. The N-dimension characteristic spacing is regarded to formulate every dataset to contain value point of every characteristic as a particular form of coordinate. As a result, categorization is conducted by locating the hyper plane, which provides the difference between the relevant classes effectively [9]. Being the coordinates of the relevant observations, which are located on the border lines, the support vectors are divided by the vector machine that specifies the decision function in the relevant support vectors. The support vector machine is centered on the framework of the decision plane, which defines the decision variations. The decision planes are one of the frameworks, which separate different forms of objects denoted with various category memberships. A schematic sample is indicated in the image below, which shows that the object belongs to the red or the green class. The line separating the objects is the boundary that defines the right area of the green objects and the left area of red objects. An additional form of object (the white circle) on the right hand side is classified as green and will be red if classified in the left side.

The above Fig. 7.6 signals a liner classifier, which separates different sets of objects in the various color groupings (Red or Green) using a line. Majority of classification tasks are not simple but often complex in structure and necessitate optimal separations such as effectively classifying objects based on examples, i.e., the train instances.

In the image below, in comparison to the prior schematic image, it is evident that there should be complete separate red and green objects separated by a curved boundary line. The categorization of tasks centered on the formation of the boundary

Fig. 7.6 Signals liner
classifier

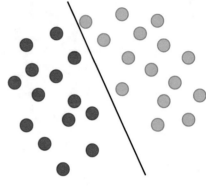

Fig. 7.7 SVM classifier

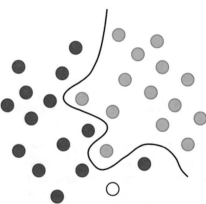

lines to differentiate objects of various category memberships are called hyper plane classifiers. The SVMs is in particular obliged to handle the problem.

The Fig. 7.7 illustrated below indicates the basic assumption over the SVMs, whereby the original object is denoted on the left area of the mapped schematic and regrouped using a various mathematical set of functions called the kernel. The object regrouping process is referred as mapping or transformation, which in a new setting; mapped images on the right area of the schematic can be separated. Other than drawing the complex curve on the left area schematic, finding the optimal boundary line to separate the red and green objects is critical Fig. 7.8.

7.6.3   Naïve Bayes

The Naïve Bayes Algorithm is one of the supervised learning methodologies whereby the speculations of every attribute belongs to every class is relative for calculation. This form of algorithm denotes that the calculation of every attribution belongs to a certain class value and is independent to other forms of attributes.

Fig. 7.8 Optimal boundary line classifier

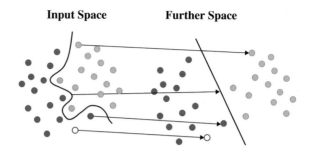

Input Space   **Further Space**

When the valuation of attributes is determined, the probability class valuation is known as the conditional probability. The dataset instances probability is denoted through the multiplication of attributes and conditional probabilities. Thus, the predictions may be calculated using every class instance and probabilities through a selection of the greatest probability value class. This classifier and theorem, name after Thomas Bayes, was introduced and developed by Pierre Laplace who was a publisher of modern Bayes equations in early 1812. Generally, the Bayes theory explains the event probability considering the previous assumptions of relevant conditions of a particular event. Thus, this theorem fits effectively in Supervised Machine Learning since it is a critical role of machine learning to make accurate predictions of the future considering prior knowledge and experience [10]. A mathematical approach of the Bayes theory is written as follows:

$$P\left(\frac{M}{N}\right) = P\left(\frac{N}{M}\right) * \frac{P(M)}{P(N)} \tag{7.4}$$

7.6.4 Random Forest

This SML classifier was proposed by Breiman in 2001 as a collaborative methodology that operates centered on proximity research. This theorem is based on the decision tree, which makes conquer and standard divide approach to enhance the calculation performance. The fundamental rule governing the random forest is the effective and firm learning cluster formed by the collection of weaker learners appropriate to the disjunction hypothesis [11]. One of the critical merits of the random forest is its utility during regression and classification of problems that form various preceding supervised machine learning systems. The image below shows the appearance of the Random Forest with two trees.

The RF classifier composes of hyper-parameters same as the bagging and decision tree classifiers. The fundamental aspect of this classifier is that they are not

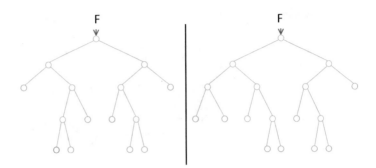

Fig. 7.9 Random Forest with two trees

composed of interlinked decision trees with bagging classifiers, which necessitate the utility of classifier RF in Fig. 7.9. The RF also deals with regression roles through the application of the RF regressor, which adds on randomness of frameworks while developing the trees. Other than investing fundamental features through the split of nodes, the model considers effective features over the random feature subsets [12]. As a result, it leads to diversity of the overall results of effective models whereby the RF is composed of subsets and feature considering through algorithms in split nodes. The decision tree may be made more random through the addition of more random threshold for every feature instead of looking for effective thresholds.

7.6.5 *Intrusion Detection Datasets*

The standard interruption identification dataset KDDCUP9923 comprises of repetitive records. Thus, it prompts unjustifiable effects of the machine learning algorithms. Therefore, the regulated machine learning algorithms are tried NSL-KDD24 dataset which is the propelled form of the KDDCUP99 interruption recognition dataset. It comprises of 42 elements and 4 recreated attacks. The denial of service attack (DoS) which include: Over use of data transmission or non-accessibility of the framework assets prompting the DoS attacks. For instance: Teardrop, Smurf, and Neptune.

The Fig. 7.10 shows methodology of Client to Root (U2R) attack at first aggressor get to typical client account, later access the root by misusing the vulnerabilities of the framework. Precedents: Perl, Load Module, and Discharge assaults. Test Assault: Has an entrance to whole system data before presenting an assault. Precedents: ipsweep, nmap assaults. Root to Neighborhood (R2L) Assault: By abusing a portion of the vulnerabilities of the system aggressor increases nearby access by sending bundles on a remote machine. Models: imap, surmise secret word, and ftp-compose assaults.

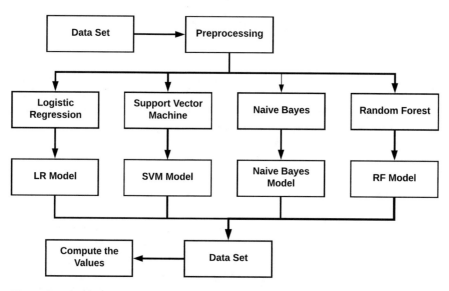

Fig. 7.10 Methodology

7.6.6 Approach

The system utilized is appeared in the Fig. 7.10. In pre-preparing step, all the information which are in printed frame are changed over to numerical shape. Pre-handled information is isolated as testing information and preparing information. The models are manufactured utilizing Strategic Relapse, Gaussian Guileless Bayes, Bolster Vector Machine, and Irregular Backwoods classifiers. These models are utilized for anticipating the names of the test information. Real marks and anticipated names are analyzed. Exactness, Genuine Positive Rate (TPR), and False Positive Rate (FPR) are registered [13]. In light of these parameters execution of the models are looked at.

Following advances are utilized to manufacture the models.

1. Dataset pre-processing
2. Division of data into testing and training data

Structuring the classifier models based on training data utilized in:

- The Guassian Naïve Bayes
- Logistic regression aiding vector machines
- The use of random forest

7.7   The Grouping of Regulated Algorithms

As indicated by [18], the administered machine learning calculations which claims more with order incorporates the accompanying: Straight Classifiers, Strategic Relapse, Guileless Bayes Classifier, Perceptron, Bolster Vector Machine; Quadratic Classifiers, K-Means Grouping, Boosting, Choice Tree, Arbitrary Backwoods (RF); Neural systems, Bayesian Systems, etc [14].

7.7.1   Linear Classifiers

Direct models for arrangement separate the data vectors into classes utilizing straight (hyper plane) choice limits. The objective of arrangement in straight classifiers in machine learning is to bunch things that have comparative component esteems, into gatherings, expressed that a direct classifier accomplishes this objective by settling on a grouping choice dependent on the estimation of the straight blend of the highlights. A straight classifier is frequently utilized in circumstances where the speed of grouping is an issue since it is evaluated the quickest classifier. Also, direct classifiers regularly work extremely well when the quantity of measurements is expansive, as in report characterization, where every component is ordinarily the quantity word a record. The rate of union among informational index factors anyway relies upon the edge. Generally, the edge measures how directly detachable a dataset is, and subsequently that it is so natural to take care of a given grouping issue.

7.7.2   Logistic Regression

This is an arrangement work that utilizes class for building and uses a solitary multinomial calculated relapse show with a solitary estimator. Strategic relapse typically states where the limit between the classes exists, additionally expresses the class probabilities rely upon separation from the limit, in an explicit methodology. This move towards the limits (0 and 1) all the more quickly when informational index is bigger. These announcements about probabilities which make strategic relapse something other than a classifier. It makes more grounded, progressively definite forecasts, and can be fit in an unexpected way; however, those solid expectations could not be right. Strategic relapse is a way to deal with forecast, similar to Conventional Minimum Squares (OLS) relapse. Be that as it may, with calculated relapse, expectation results in a dichotomous result. Calculated relapse is a standout among the most generally utilized apparatuses for connected insights and discrete information investigation. Strategic relapse is direct interpolation.

7.7.3   The Naïve Bayesian Networking System

These are extremely straightforward Bayesian systems which are made out of coordinated non-cyclic charts with just a single parent (speaking to the in secret hub) and a few youngsters (comparing to watch hubs) with a solid supposition of autonomy among kid hubs with regard to their parent [15]. Thus, the freedom demonstrates (Innocent Bayes) depends on evaluating. Bayes classifiers are typically less exact than other increasingly [17] complex learning calculations, (e.g., ANNs). However, played out an expansive scale correlation of the gullible Bayes classifier with cutting edge calculations for choice tree enlistment, occasion-based learning, and principle acceptance on standard benchmark datasets, and observed it to be at times better than the other learning plans, even on datasets with significant element conditions. Bayes classifier has attribute independence issue which was tended to with arrived at the midpoint of One-Reliance Estimators.

7.7.4   The Multilayer Perception

This is a classifier in which the heaps of the framework are found by dealing with a quadratic programming issue with straight restrictions, rather than by settling a non-raised, unconstrained minimization issue as in standard neural framework getting ready. Other well-known calculations depend on the idea of perception. Perception calculation is utilized for gaining from a group of preparing examples by running the calculation over and over again through the preparation set until the point when it finds an expectation vector which is right on the majority of the preparation set. This expectation rule is then utilized for foreseeing the marks on the test set [18].

7.7.5   Supporting Vector Machines (SVRs)

These are the latest administered machine learning procedures. Support Vector Machine (SVM) models are closely related to traditional multilayer perception neural networks. SVMs spin around the idea of a margin [19] on either side of a hyper plane that isolates two information classes. Boosting the edge and thusly making the greatest possible detachment between the secluding hyper plane and the events on either side of it has been exhibited to diminish an upper bound on the ordinary theory mistakes.

7.7.6 K-Implies

K-implies is one of the slightest troublesome unsupervised learning estimations that deal with the striking gathering issue. The technique seeks after an essential and straightforward way to deal with portray a given educational record through an explicit number of gatherings (acknowledge k clusters) settled from the earlier. K-Means calculation is be utilized when marked information isn't accessible. Given weak learning calculation that can reliably discover classifiers at any rate somewhat superior to anything irregular, state, exactness 55%, with adequate information, a boosting calculation can provably develop single classifier with high precision, state, 99%.

7.7.7 Decision Tree

Decision Trees (DT) are trees that arrange models by orchestrating them reliant on feature regards. Each center point in a decision tree addresses a component in an event to be requested, and each branch addresses a regard that the center can acknowledge. Events are start at the root center point and masterminded reliant on their component regards. Decision tree learning, utilized in information mining and machine learning, utilizes a choice tree as a prescient model which maps perceptions around a thing to decisions about the thing's objective esteem. Increasingly unmistakable names for such tree models are arrangement trees or relapse trees [16]. Decision tree classifiers as a rule utilize post-pruning methods that assess the execution of choice trees, as they are pruned by utilizing an approval set. Any hub can be expelled and doled out the most widely recognized class of the preparation occurrences that are arranged to it.

7.7.8 Neural Systems

Neural Systems (NS) that can really play out various relapse as well as grouping errands without a moment's delay, albeit generally each system performs just a single. In most by far of cases, in this manner, the system will have a solitary yield variable, in spite of the fact that on account of many-state order issues, this may be compared to various yield units (the post-handling stage deals with the mapping from yield units to yield variables). Artificial Neural System (ANS) relies on three principal viewpoints, information and initiation elements of the unit, organize design, and the heaviness of each info association. Given that the initial two perspectives are settled, the conduct of the ANS is characterized by the present estimations of the loads. The loads of the net to be prepared are at first set to arbitrary qualities, and after that occasions of the preparation set are over and over presented to the net. The qualities for the contribution of an occasion are set on the info units and the yield of the net is contrasted and the ideal yield for this case. At that point, every one of the loads in the net are balanced marginally towards the path that would bring the yield estimations of the net nearer to the qualities for the ideal yield. There are a few calculations with which a system can be prepared.

Bayesian Networking System: A Bayesian Framework is a graphical model for probability associations among a great deal of components. Bayesian frameworks are the most prominent specialists of verifiable learning counts. The most intriguing component of BNs, stood out from decision trees or neural frameworks, is certainly the probability of thinking about prior information about a given issue, with respect to helper associations among its features. An issue of BN classifiers is that they are not fitting for datasets with various features. This prior bent, or space data, about the structure of a Bayesian framework can take the going with structures:

- Declaration of the node being a root node, for example, lacking a parent
- Declaration of the node being a leaf node, for example, lacking children
- Declaration of the node being a direct effect or direct cause of a dissimilar node
- Declaration that a node is not directly associated to another node
- Proclaiming two independent nodes provided a set of conditions
- Provision of fractional nodes in order, that is, showing that a node shows up earlier compared to a different ordering node
- Provision of an absolute ordering node

7.8 Elements of a Learning Machine Algorithm

Regulated machine learning systems are material in various areas. Various Machine Learning (ML) application can be found in [1]. By and large, SVMs and neural systems will in general perform much better when managing multi-measurements and consistent highlights. Then again, rationale-based frameworks will in general perform better when managing discrete/absolute highlights. There is general understanding that k-NN is extremely delicate to insignificant highlights: this trademark can be clarified by the manner in which the calculation works. In addition, the nearness of superfluous highlights can make neural system preparing exceptionally wasteful, even unfeasible. Most choice tree calculations can't perform well with issues that require corner-to-corner parceling. The division of the case space is symmetrical to the pivot of one variable and parallel to every other hatchet. In this manner, the subsequent districts in the wake of apportioning are all hyper rectangles.

The ANNs and the SVMs perform well when multi-co linearity is available and a nonlinear relationship exists between the info and yield highlights. Credulous Bayes (NB) requires little storage room amid both the preparation and order organizes: the strict least is the memory expected to store the earlier and contingent probabilities. The fundamental kNN calculation utilizes a lot of storage room for the preparation stage, and its execution space is in any event as large as its preparation space. Actually, for all non-languid students, execution space is generally a lot littler than preparing space since the subsequent classifier is normally a profoundly dense synopsis of the information. In addition, Innocent Bayes and the kNN can be effectively utilized as gradual students though rule calculations can't. Guileless Bayes is normally hearty to missing qualities since these are just disregarded in registering probabilities and thus have no effect on a ultimate conclusion. Actually, kNN and neural systems require finish records to do their work.

At long last, Choice Trees and NB for the most part have diverse operational profiles, when one is extremely precise the other is not and the other way around. Unexpectedly, choice trees and guideline classifiers have a comparative operational profile. SVM and ANN have likewise a comparative operational profile. No single learning calculation can consistently outflank different calculations over all datasets. Distinctive informational indexes with various sorts of factors and the quantity of examples decide the kind of calculation that will perform well. There is no single learning calculation that will outflank different calculations dependent on all informational collections as indicated by no free lunch hypothesis. Table 7.1 shows the similar investigation of different learning calculations.

7.9   Summary and Future Work

In conclusion, the grouping in supervised machine learning necessitates concise adjustments of different parameters and consequently with a sizeable number of sets of data. The main necessities required for the structuring of such models are proper classification and accurate precision exclusive of time. Hence excellent learning process for a good learning algorithm of a certain set of data does not always assure precision but also promises the accurateness of a dissimilar dataset whose composition are totally diverse from the others. Nonetheless, the main key problem when managing the ML characterization isn't whether the algorithms are better than others, however under which conditions a specific strategy may considerably do better than others based on a given application predicament. On the other hand, meta-learning is going towards this path, attempting to discover functions which delineate to the performance of the algorithms. Concluding the meta-learning systems are in use of vast number of properties known as the meta-attributes which correspond directly to each learning undertakings, and investigations for the relationships amid these characteristics and execution of the learning algorithms. A number of distinctive qualities of learning errands include: number of occasions, extent of definite out traits, proportion of the missing figures, entropy of classes, and so forth which highly gave a broad rundown of data and factual evaluation for each dataset.

Table 7.1 Comparison of learning algorithms (***indicate high performance while * show worst performance)

	Decision trees	Neural networks	Naïve Bayes	kNN	SVM	Rule-learners
Accuracy in general	**	***	*	**	****	**
Speed of learning with respect to number of attributes and the number of instances	***	*	****	****	*	**
Speed of classification	****	****	****	*	****	****
Tolerance to missing values	***	*	****	*	**	**
Tolerance to irrelevant attributes	***	*	**	**	****	**
Tolerance to redundant attributes	**	**	*	**	***	**
Tolerance to highly interdependent attributes (e.g., parity problems)	**	***	*	*	***	**
Dealing with discrete/binary/continuous attributes	****	*** (not discrete)	*** (not continuous)	*** (not directly discrete)	** (not discrete)	*** (not directly continuous)
Tolerance to noise	**	**	***	*	**	*
Dealing with danger of overfilling	**	*	***	***	**	**
Attempts for incremental teaming	**	***	****	****	**	*
Explanation ability/ transparency of knowledge/ classifications	****	***	****	****	**	*
Model parameter handling	***	*	****	***	*	***

Upon comprehension of the strengths and the limits of every technique, thus the likelihood of incorporating at least two algorithms for handling a task ought to be examined. Thus, the primary goal is using strengths for one strategy to supplement the shortcomings of the other. In the event that we are just inspired by the most ideal characterization accuracy, it is difficult locating a solitary classifier that executes and a decent assembly of classifiers. The RF NB and the SVM algorithm machine learning also conveys high accuracy and precision paying little attention to the number of properties and data occurrences. This examination demonstrates that the duration of assembling a model is a factor on the one hand; with a high accuracy with the kappa measurement while the MAE is considered as a different issue.

Consequently, the ML algorithms are in need of high accuracy, while fewer miscalculations should have administered prescient machine learning. The work suggests that for an expansive set of data, a dispersed handling condition ought to be well thought-out. As a result, it will make space for an abnormal state of connection among the factors that will at last make the output of the model progressively proficient.

References

1. Zhang, N.: Semi-supervised extreme learning machine with wavelet kernel. Int. J. Collab. Intell. **1**(4), 298 (2016)
2. Praveena, M., Jaiganesh, V.: A literature review on supervised machine learning algorithms and boosting process. Int. J. Comput. Appl. **169**(8), 32–35 (2017)
3. Bostik, O., Klecka, J.: Recognition of CAPTCHA characters by supervised machine learning algorithms. IFAC-PapersOnLine. **51**(6), 208–213 (2018)
4. Drotár, P., Smékal, Z.: Comparative study of machine learning techniques for supervised classification of biomedical data. Acta Electrotech. Inform. **14**(3), 5–10 (2014)
5. Suganya, M., Anandakumar, H.: Handover based spectrum allocation in cognitive radio networks. In: 2013 International Conference on Green Computing, Communication and Conservation of Energy (ICGCE), Chennai, pp. 215–219 (2013)
6. Anandakumar, H., Umamaheswari, K.: Energy efficient network selection using 802.16g based GSM technology. J. Comput. Sci. **10**(5), 745–754 (2014)
7. Matuszyk, P., Spiliopoulou, M.: Stream-based semi-supervised learning for recommender systems. Mach. Learn. **106**(6), 771–798 (2017)
8. Belkin, M., Niyogi, P.: Semi-supervised learning on riemannian manifolds. Mach. Learn. **56** (1–3), 209–239 (2004)
9. Sarkar, S., Soundararajan, P.: Supervised learning of large perceptual organization: graph spectral partitioning and learning automata. IEEE Trans. Pattern Anal. Mach. Intell. **22**(5), 504–525 (2000)
10. Ma, J., Wen, Y., Yang, L.: Lagrangian supervised and semi-supervised extreme learning machine. Appl. Intell. **49**(2), 303–318 (2019)
11. Chen, K., Shihai, W.: Semi-supervised learning via regularized boosting working on multiple semi-supervised assumptions. IEEE Trans. Pattern Anal. Mach. Intell. **33**(1), 129–143 (2011)
12. Huang, R., Zhou, P., Zhang, L.: A LDA-based approach for semi-supervised document clustering. Int. J. Mach. Learn. Comput. **4**(4), 313–318 (2014)
13. Subramanya, A., Talukdar, P.: Graph-based semi-supervised learning. Synth. Lect. Artif. Intell. Mach. Learn. **8**(4), 1–125 (2014)
14. Krogel, M., Scheffer, T.: Multi-relational learning, text mining, and semi-supervised learning for functional genomics. Mach. Learn. **57**(12), 61–81 (2004)
15. Iosifidis, A.: Extreme learning machine based supervised subspace learning. Neurocomputing. **167**, 158–164 (2015)
16. Nishii, R.: Supervised image classification based on statistical machine learning. SPIE Newsroom. (2007)
17. Sądel, B., Śnieżyński, B.: Online supervised learning approach for machine scheduling. Schedae Informaticae. **25**, 165–176 (2017)
18. Anandakumar, H., Umamaheswari, K.: An efficient optimized handover in cognitive radio networks using cooperative spectrum sensing. Intell. Autom. Soft Comput. 1–8 (2017)
19. Anandakumar, H., Arulmurugan, R., C. C. Onn.: Computational intelligence and sustainable systems. In: EAI/Springer Innovations in Communication and Computing (2019)

Chapter 8
Green Wireless Communications Via Cognitive Handover

8.1 Introduction

The most principal analysis of the Green Wireless Communication (GWC) literature works has been the sensational advancement in remote and adaptable CR. The environmental change is one of the certain compelling issues in the modern-day technology. The explanation behind expanded greenhouse gases (GHG), fundamentally carbon dioxide (CO_2), is the expanded vitality utilization, which results in the increased volume of gases as argued in [1]. Disasters like hurricanes, surges, and changes in the sea levels are credited to the CO_2 filled greenhouse impact. It is unsurprising that amid the most recent 30 years the CO_2 emanations have increased by 73%. India is positioned the fifth among the nations in the rundown of overall GHG emanation, with the USA and China contributing around multiple times discharge than that of India. The Kyoto Protocol of 1997, which was marked by more than 160 nations, including India, approaches all nations to decrease their emanations of greenhouse gases by 5%, from the 1990 dimension, constantly 2012. Numerous administrations around the globe, including India, have found a way to diminish vitality utilization and emanations. India is focused on decreasing carbon force by 20–25% somewhere in the range of 2005 and 2020. The data and correspondences innovation industry alone records about 2% of 860 million tons of the world's greenhouse gas emanations [2].

The principle causative segments in the ICT business incorporate the vitality prerequisites of PCs and screens about 40%, server farms about 23%, and versatile media transmission contributes about 24% of the aggregate discharges. Contrasted with alternate areas, for example, transport, development, and vitality generation, the ICT division is relatively vitality lean with media transmission contributing simply 0.7% or around 230 million tons of greenhouse gas discharges. The test for the administration, media transmission, specialist co-ops, and telecom gear makers is to seek after development in the media transmission division while guaranteeing

© Springer Nature Switzerland AG 2019   155
A. Haldorai, U. Kandaswamy, *Intelligent Spectrum Handovers in Cognitive Radio Networks*, EAI/Springer Innovations in Communication and Computing, https://doi.org/10.1007/978-3-030-15416-5_8

that the 2% of worldwide outflows do not fundamentally increment over the coming years. A run of the mill correspondences organization spends almost 1% of its incomes on vitality, which for huge administrators may add up to many crores of rupees. An Earth-wide temperature boost is a significant issue, and Intergovernmental board on environmental change (IPCC) has detailed that the outflows of greenhouse gases (GHG) must be divided by the center of this century.

Nevertheless, the vast majority of the exploration endeavors underscores headway in innovation and overlooks the antagonistic impact of innovation on nature. In this way, more significance ought to be given to building up the productive advancements. Media transmission ventures are developing at a quick rate. These days, there are in excess of four billion mobile phone supporters on the planet. The quick development of endorsers energizes quick up degree in advancements. While following the advancement of new advances, the quantity of base stations will likewise increment [3]. Various base stations prompted more utilization of vitality. The information demonstrates that the arrangement of portable broadband system, for example, Long Term Evolution (LTE) is relied upon to happen on the highest point of existing 2G and 3G systems with an expansion in around 25% in the quantity of base stations. This will prompt more discharge of carbon dioxide in the environment. Specialist organizations are likewise confronting the issue of vitality use.

Physical association must be made to the client. Subsequently, the power utilization is profoundly associated the supporters' count, and customers' power is an enduring motivating force for an explicit advancement. Flexible access frameworks are proposed to cover an explicit zone and interconnect the customers here. Therefore, the power usage per customer is dependent on the customer thickness of the anchored district. In case we expect a customer thickness of 300 customers/km, we get a for every customer control usage of generally 16.5 Watt/customer for adaptable WiMAX, 5.0 Watt per customer for settled WiMAX, and 6.0 Watt per customer for UMTS. This is higher than the power usage for settled line get to frameworks. We need to consider in any case for adaptable access organizes passage to be considered. The customer ought to be adaptable and the client device is as such ordinarily a phone or a USB dongle.

These contraptions are redesigned for minimal usage of power to achieve long self-rule. While differentiating the power use for admittance framework and CPE in this circumstance, the power use per customer is equal between settled line propels and flexible advancements. When we, regardless, consider only a vast segment of the customer thickness (or two versatile access systems from contending administrators covering the region), the per-client control utilization of the portable access arrange promptly copies. This unique ends the power utilization of portable access systems. The information demonstrates that with respect to the CO_2 emanation per supporter every year, 4.3 kg will be radiated from base station rather than 8.1 kg for client hardware at the assembling stage, at the capacity, is 9 kg for base station rather than 2.6 kg for the client gear. In the event that we need to diminish the framework control cost of a continuous cell framework, more undertakings ought to be put on the vitality sparing of the base station side.

8.2 Green Wireless Communication

The term green wireless communication can be characterized as the innovation, which utilizes the combination of vitality productive systems at various stages to limit the unfriendly impacts of innovation on nature. Developing media transmission framework requires an expanding measure of power to control it. India right now has in excess of 310,000 PDA towers, which expend around two billion liters of diesel for each year. The move from diesel to sun based and other number wellsprings of vitality will result in a decrease of five million tons of CO_2 emanations and in addition, an investment fund of \$1.4 billion in working costs for media transmission in tower organizations. Moving to sustainable power sources can create a huge number of carbon credits that could counterbalance the opex on their towers. Furthermore, sparing in the vitality bill would additionally decrease the working cost. Green remote correspondence has numerous features. It tends to be characterized extensively as far as the greening of media transmission systems, green telecom gear maker, and safe media transmission squander transfer.

Specialized part managing IT administrations is assuming a main job in greenhouse gas discharges worldwide and is in charge of around 2–2.5% of destructive greenhouse emanation with the developing administration requests. The aggregate vitality utilization of remote correspondence frameworks and systems and the web is about over 2.5% of the aggregate vitality use nowadays and the sum will clearly build even more immensely later on. Hence, keeping up vitality proficiency is an ongoing pattern to be pursued for planning future remote correspondence hardware. Vitality use in remote systems at both foundation level and among battery worked remote hubs is one noteworthy concerned issue in modern or specialized associations. Accordingly, the exertion is made to determine the issue by proposing instruments and hardware that enhance the vitality productivity of correspondence systems. ICT causes 2% of overall CO_2 outflows.

The idea of green remote correspondence expects a wide number of outlook changing propelled specialized methodologies, for example, enhancement strategies, vitality proficient system design, and productive systems administration conventions. These can better transmission and strategies range (as stressed by cognitive radio, that is, auxiliary clients could utilize the range without meddling with the exercises of essential clients) actualizing the possibility of a green range. Expanded vitality utilization for future necessities will prompt the expansion in CO_2 discharges, which prompts climatic changes, rising toxin content, cataclysmic events, and ocean level adjustments, in this way irritating the nature's environmental equalization. Most divisions add up to emanations incorporate PCs and Monitors (40%), Data focuses (23%) and Mobile broadcast communications (24%). Subsequently, the need of great importance is to utilize eco-accommodating inexhaustible assets of vitality and decrease in carbon discharges to lessen the dimension of contaminations in nature. To keep up biological concordance to guarantee parity in nature and avert carbon outflows into the air, advancement of green mechanical improvement is a noteworthy perspective to give vitality effectiveness in remote correspondence, prompting monetary creation levels without trading off with the nature of administration.

Green remote correspondence Green Wireless Communication is a clarification in itself. It utilizes eco-accommodating procedures and methodologies in correspondence systems to limit wastage of vitality and impacts of mechanical headways on common equalization. Increment in media transmission requests and gear requires the expanding measure of power to work it. Therefore, there is a requirement for moving from non-sustainable wellsprings of vitality to condition inviting sustainable power sources. It will result in a decrease of CO_2 discharges because of open copying of petroleum derivatives and in addition sparing of working costs for media transmission organizations.

8.3   Green Telecommunication Networks

In media transmission systems, greening advances the possibility of vitality productive remote correspondence utilizing eco-accommodating vitality choices and innovations like sustainable power source assets.

8.3.1   Green Manufacturing

The way towards assembling ought to be continued while dealing with its consequences for condition by utilizing eco-accommodating and vitality effective innovation. This also includes hardware and assembling gear, recycling and appropriate transfer of electronic and mechanical squanders, least utilization of poisonous materials like lead and mercury by enterprises that are hard to break down causing soil contamination, forbidding the utilization of plastics, and decrease of unsafe radio outflow into environment to maintain a strategic distance from air contamination. All means making nature hurt ought to be supplanted by elective techniques.

8.3.2   Disposal of Waste

Transfer of electronic and mechanical waste (e-squander transfer) gear ought to be done in an eco-accommodating way with the goal that any perilous substance utilized amid generation process ought not to contaminate nature's air, water, or soil. Every one of the sorts of contamination caused because of mechanical advancement ought to be limited to achieve reasonable improvement, that is, improvement should happen without exasperating the nature's parity, safeguarding the earth, and without trading off with the necessities of who and what is to come. Preservation of assets ought to be done as such that they could be accessible for use by whom and what is to come.

8.4   Techniques to Reduce Carbon Footprints

Ongoing examination has demonstrated that a discernible decrease in the carbon discharges because of media transmission industry is conceivable through various preventive measures by contemplating the mother Earth. The Kyoto convention has expressed those noteworthy gases that are adding to the unnatural weather change impact and are known as GHG (greenhouse gases). This consists of methane, carbon dioxide, sulfur hexafluoride, PFCs, and HFCs, which consist of distinctive dimensions of influencing the Earth and cause collection of carbon into layers of the climate. Beginning from the assembling of electronic segments utilized in the correspondence system to the different advances associated with structuring the system foundation hardware and to the task life expectancy of the system, including at last system disappointment and e-squander transfer, there are sure exercises that produce greenhouse gases.

These are specific cause for harmful effects on nature and human wellbeing which can be controlled utilizing following causes:

- Watchful cell arranging with the goal that a wide inclusion can be accomplished by utilizing a lesser number of receiving wires.
- Framework sharing ought to be permitted so as to limit perilous effects on nature.
- Supplant climate control systems that discharge CFCs into the air with elective cooling techniques.
- Utilizing HFC free cooling frameworks.
- Open air base stations ought to be favored.
- Structuring vitality effective instruments and hardware that can work well expending less power and utilization of sustainable wellsprings of vitality like sun oriented, tidal, hydral, and wind vitality.

The cognitive radio (CR) idea is developing as a standout among the most encouraging methods in remote interchanges for tending to the range shortage by dynamic range. The CR transmission process is commonly made out of two fundamental stages:

- The range detecting stage, where CR clients endeavor to recognize the unused range (alluded to as a range opening) for instance in television recurrence sections
- The transformation section of cognitive networks whose clienteles transfer data across spectrum gaps denoted

Consequently, broad research endeavors have been committed to the CR idea, the majority of which are centered around range detecting, access, and designation and sharing, going for enhancing the otherworldly proficiency, instead of the vitality productivity. The channel limit is directly corresponding to the data transfer capacity, according to the logarithms into the transmission controls. This will be presented as information transmission watches out for limitlessness; the limit turns out to be additionally an expanding capacity of the power. This infers that given a specific target limit, the utilization of extra otherworldly data transmission can help diminish the transmission control to keep up a given channel limit, exhibiting the exemplary tradeoff between the power- and transfer speed effectiveness. Half of the aggregate vitality can be spared, if versatile system administrators utilize CR strategies for progressively dealing with their authorized range groups and completely misusing the accessible radio assets for vitality proficiency enhancements. As an outcome, CR is definitely getting to be one of the key procedures of future green remote systems.

Participation (additionally named as agreeable interchanges) is likewise all around perceived as a hot research subject in remote correspondences. Helpful interchanges enable the circulated system hubs to help each other for amplifying their particular advantages, which gives another viewpoint on remote system enhancement. This profession was spearheaded in [4], where a few agreeable handing-off conventions (i.e., settled handing-off, specific transferring, and gradual handing-off) were contemplated as far as both the reachable rate and blackout likelihood. In [5], the work of room time coding in agreeable remote systems was additionally inspected, appearing noteworthy enhancement in blackout execution through client participation. It was shown in [6] that agreeable strategies are equipped for enhancing the transmission dependability with the guide of their spatial assorted variety gain or potentially expanding the framework throughput through asset collection.

At the season of composing, helpful correspondence designs have just been received in a few remote systems administration gauges, e.g., IEEE 802.16j and the long haul development (LTE)—propelled frameworks. Notwithstanding enhancing unwavering quality and throughput, agreeable interchanges likewise offer potential vitality investment funds in remote systems. Considering this case, the systems in the cells, when the client is perpendicular to the cell edge, it is fundamental to transmit maximum control edge to maintain the objectivity of beneficial nature, which is necessitude in the uplinks (ULs). This can critically diminish the vitality of battery, which might further lead to critical co-channeling obstructions. For this situation, choosing a fitting hand-off to help the cell edge client's information transmission is prepared to do viably diminishing its capacity utilization. For instance, having an ideal unravel and forward transfer in the MS and BS due to decreasing of requires power and requesting dimension. In this manner, it is of high handy enthusiasm to investigate agreeable strategies for vitality reserve funds in cell systems, particularly for the cell edge clients [15].

Generally, the present remote gadgets (e.g., cell phones) are furnished with numerous system get to interfaces, for example, Bluetooth, Wi-Fi, and LTE, where the diverse remote access systems include distinctive radio qualities as far as their inclusion territory and vitality utilization. To be explicit, both Bluetooth and Wi-Fi give neighborhood at low vitality utilization, though LTE offers a more extensive inclusion, however tragically at a higher vitality utilization. Persuaded by the perception that diverse remote systems supplement one another, we have explored the participation between various heterogeneous remote systems (e.g., Bluetooth and LTE) in [7], which might be alluded to as system collaboration. All the more explicitly, we have inspected the utilization of Bluetooth to help LTE transmissions for enhancing the vitality proficiency of cell correspondences. It was presented in [8] that given the explicit target information rate and blackout likelihood prerequisites, participation has the capability of altogether lessening the aggregate vitality utilization of cell transmissions, particularly when the versatile stations (MSs) are at the cell's edge [9].

In the proposed system, cell gadgets are first permitted to recognize range gaps through range detecting. At that point, arrange collaboration is conjured for effectively misusing the accessible range openings for green correspondences. The advantages of comprehension have been researched for cell correspondences in [10] where the overlay approach was conjured for range sharing between a TV communicate framework and a phone framework. The overlay-based range sharing was appeared to be gainful as far as expanding the limit of both the downlink and uplink of cell correspondences. In [11], the cognitive system was likewise examined in cell systems for shrewdly allotting system assets to relieve the impacts of co-channel impedance. Moreover, in [12], the collaboration among numerous radio access innovations (RAT) was inspected concerning heterogeneous remote systems. Despite the fact that the cognizance and participation strategies have been contemplated in cell systems, these exploration endeavors were centered around the improvement of the system limit and inclusion, rather than enhancing the remote vitality proficiency.

Even more as of late, in [13], vitality gathering (EH) was conjured for diminishing the power utilization of remote sensor systems. As opposed to the system limit enhancement techniques for [10–12], we plan to abuse the joint advantages of cognizance and participation for upgrading the vitality productivity of cell correspondences. The principle commitments of this paper are abridged as pursues. Right off the bat, we propose a cognitive system participation structure, where distinctive remote systems cognitively coordinate with one another for enhancing the feasible vitality effectiveness of remote correspondences. Furthermore, we present the joint discernment and collaboration helped cell interchanges theory and evaluate its vitality sparing advantages contrasted with both the conventional direct transmission and to the unadulterated comprehension and unadulterated participation. At last, we uncover the tradeoff between the vitality effectiveness and blackout execution, exhibiting that the vitality productivity of cell interchanges enhances, as the blackout execution corrupts.

Whatever is left of this paper is composed as pursues. The next segment surveys the range detecting approaches considered for distinguishing range openings in a vitality effective way. In the accompanying segment, we examine the cognitive system collaboration structure summoned for productively using the range gaps for vitality decrease in cell systems. Next, Section 8.5 exhibits the contextual analysis of a vitality productive cell organize plan and demonstrates that the vitality proficiency of cell interchanges can be fundamentally enhanced by abusing the simultaneous utilization of cognizance and collaboration.

8.5   Spectrum Range Sensing Relying on Cognition

This segment is fundamentally centered on the distinguishing proof of range openings by means of cognitive functionalities. As per the Shannon limit of a band limited boisterous channel, there is a tradeoff among transfer speed and power-proficiency, inferring that expanding the transmission data transfer capacity is fit for lessening the transmit control without debasing the channel limit. In the meantime, the applicable investigations have demonstrated that the control of the authorized range normally differs in the scope of 15–85%; in this manner, an extensive part of the range is under-used. Specifically, some recurrence groups, for example, TV groups, are largely vacant by authorized clients at certain geographic areas and specific time interims. Subsequently, it is helpful [16] to recognize and misuse the unused range in these groups for cell correspondences to lessen vitality utilization, while fulfilling both the throughput and QoS prerequisites. There are two run of the mill range openings, specifically the fleeting and spatial range openings. To be explicit, whenever authorized (or essential) clients do not involve an otherworldly band at a specific time, we can incidentally reuse it for unlicensed (or auxiliary) clients, which are alluded to as a fleeting phantom chance. Conversely, if the licensed and unlicensed clients are adequately far from one another forced when they transmit over a similar recurrence band, the unlicensed clients can completely reuse the authorized clients' ghostly band, which might be named as a spatial-otherworldly chance [17].

8.6   Techniques from Spectrum Distinguishing

The distinguishing of the spectrum is fundamental for the empowerment of cognitive radio to learn, qualify, and acknowledge the status of work, for instance the status of obstruction and ranging accessibility. At the point when a specific recurrence band is identified as underutilized by the essential/authorized client at a specific time in an explicit position, the auxiliary clients can use the range, i.e., a range opportunity exists. Thus, the distinguishing of the spectrum might be performed over the space, recurrences, and during time regions. Concerning the progressive enhancements of various clients, the beam forming initiatives may apply the same recurrences and channels in regional areas. Subsequently, if an essential client (PU) is not transmitting in complete disorder, range openings can be made for optional clients in the ways, not in administration, and range detecting must likewise consider the point of landings. The essential clients can likewise utilize their relegated groups by methods for spread-range or recurrence jumping, and optional clients would then be able to transmit in a similar band at the same time without extreme interruption to the essential clients, if they receive a symmetrical code in connection to the codes embraced by the essential clients [18].

This makes range openings in code space yet requires identification of the codes utilized by the essential clients and also multipath parameters. Since identifying essential clients that are accepting information is commonly extremely troublesome, numerous investigations on range detecting have concentrated on essential transmitter recognition dependent on the nearby estimations of auxiliary clients. Spectrum detecting and channel testing to obtain ongoing range/channel data required by the cognitive MAC layer are likewise basic segments of cognitive radio systems. When all said is done, range detecting plays out the accompanying errands: (1) discovery of range openings, (2) assurance of otherworldly goals for every range gap, (3) estimation of the spatial headings of an approaching meddling sign, and (4) flag characterization. Among these, the identification of range gaps is likely the most essential assignment and is investigated through a double theory testing issue [17].

In this way, discovery of range openings on a restricted recurrence band is generally alluded to as range detecting, which identifies the nearness or nonattendance of essential clients in the basic band. Range detecting methods can be separated into two principle classes: non-agreeable/transmitter discovery and helpful identification. The approach for the location of transmitters depends on signal discovery that are transmitted from relevant frameworks using the neighboring cognitive radio perceptions and clients. Non-cooperative transmitters and discovering procedures are supportive in regions with the relevant transmitters, which are obscure to cognitive radio devices. In this manner, cognitive clients ought to depend just on the location of frail essential transmitter flags and utilize just nearby perceptions to perform range detecting. A cognitive gadget does not have finish learning of range inhabitance in its inclusion territory. As a result, it is absurd to evade hurtful obstruction with essential clients.

A cognitive client (CU) may have a decent observable pathway with an essential collector, yet will be unable to recognize the nearness of an essential transmitter (shrouded terminal) because of the shadowing wonder, which is exceptionally basic in urban/indoor conditions. Helpful discovery systems are actualized to moderate this issue. Agreeable location alludes to range detecting techniques that empower numerous cognitive radios to share their neighborhood detecting data for progressively exact essential transmitter discovery. Agreeable location can be executed in either an incorporated or a conveyed way. In the brought together strategy, a focal unit gathers detecting data from cognitive gadgets, distinguishes the accessible range groups, and communicates this data to other cognitive radios. In a circulated methodology, there is no focal hub, and the detecting data is shared among the cognitive gadgets. Disseminated discovery is simpler to actualize, does not require a spine foundation, while unified recognition is progressively precise, and can adequately relieve both multi-way blurring and shadowing impacts. The focal hub can likewise allocate an explicit load to every range detecting result to alleviate blurring wonders.

Cooperative recognition systems can be likewise named a delicate or hard blend, as per the idea of the data shared among cognitive clients. The delicate mix alludes to a helpful methodology in which every hub detects a specific recurrence band and after that sends the consequences of its measures—i.e., the vitality of the got flag—to the focal hub. On the other hand, in hard blend procedures, every hub chooses whether an essential client is available and after that reports to the focal hub just the consequences of its choice. Delicate identification is typically progressively exact and can execute full-scale assorted variety procedures, assignments received from far off hubs will, in general, be uncorrelated. Hard discovery is not as exact; however, it requires less data trade between hubs. In the event that a cognitive gadget is furnished with different reception apparatuses, modern detecting systems can be actualized, misusing spatial, time, as well as recurrence coding. Such agreeable range detecting is illustrated in detail in [14], and the creators exhibit that the likelihood of false cautions can be decreased utilizing space, time, and recurrence transmit. Transfer variety can be additionally utilized to adjust for the decreased detecting assorted arrange when a few hubs in a helpful range detecting framework can't report specifically to the focal hub (i.e., because of shadowing wonder).

For the most part, range detecting is performed by utilizing straightforward flag discovery strategies to distinguish abandoned frequencies as fast as would be prudent. In any case, these straightforward methods cannot accomplish dependable and exact detecting results in low-SNR and profound blurring situations. Different strategies have been proposed to improve the dependability and exactness of range identification including the combination of numerous nearby discovery choices and agreeable range detecting.

The determination of the most appropriate identification strategy for nearby range detecting is a noteworthy test since location procedures contrast in their execution. For instance, the vitality finder (ED) cannot identify signals with low SNR. This can be accomplished with the violent wind stationary element indicator (CSFD), however with included time and multifaceted nature. The coordinated channel (MF) is the ideal identification strategy if the PU's data is known. As opposed to the coordinated channel and violent wind stationary component indicator, nevertheless, the vitality locator requires no earlier learning of the essential client flag.

In the ongoing writing on range detecting, three distinct kinds of flag recognition techniques have been considered for the distinguishing proof of range gaps, to be specific vitality discovery, coordinated sifting, and highlight extraction. By and large, a vitality indicator amasses the vitality of the flag acquired over a recurrence band and afterward takes a gander at it to a predefined edge to pick whether the unearthly band is controlled by approved customers or not. Being express, if the totaled imperativeness is lower than as far as possible, the watched range band is regarded to be idle; something different, the band is seen as being included by the approved customer. Observing that is impossible in the ED. Promptly separate the ideal flag from both the obstruction and the clamor, henceforth it is inclined to missed location or false caution occasions activated by the impedance and commotion. Thus, the MF based identifier was anticipated as compelling methods for battling the impedance, which is viewed as the ideal locator in added substance white Gaussian clamor situations.

Table 8.1 A rationale between a matched filtered sensor, energy sensor, and an extraction detector

Techniques	Noise variance	Prior knowledge	Computation complexity	Energy consumption
Energy detector	High	Low	Low	Low
Matched filter detector	Moderate	High	Moderate	Moderate
Extraction detector	High	Moderate	High	High

Nonetheless, the MF requires some earlier learning of the essential flag to be distinguished, for example, the beat shape and balance type. As a further option, the element extraction helped locator develops as a promising detecting approach, which is able to do adequately recognizing the essential signs both from the foundation commotion and from the obstruction. This makes the FE finder strong to the foundation commotion even in amazingly low flag to-clamor proportion situations. This preferred standpoint of the FE locator anyway comes at the expense of a high computational unpredictability, since it requires an additional preparation procedure for separating the important flag highlights. Because of its high computational multifaceted nature, the FE locator devours more vitality than the ED and the MF indicator. Table 8.1 gives a synopsis of the three kinds of locators regarding their heartiness to commotion fluctuation vulnerability, the earlier information prerequisite, the computational multifaceted nature forced, and the vitality utilization.

Both the ED and in addition the MF identifier and the FE finder ordinarily function admirably in Gaussian clamor conditions, yet their recipient working qualities seriously debase in remote blurring situations. In particular, if a profound blur is experienced, the ideal flag got at an unlicensed client may turn out to be too powerless to be in any way identified by the previously mentioned ED, MF, and FE finders, in this manner causing an execution debasement. To battle and control the blurring impacts, agreeable range detecting might be conjured by enabling various clients to participate in recognizing range openings. Different clients first output the authorized range groups and after that report their free perceptions to a combination community for settling on a ultimate choice concerning the inert or occupied status of the filtered phantom groups. It has been appeared in that helpful range detecting essentially beats the regular non-cooperative methodologies.

This, in any case, comes to the detriment of extra vitality utilization, since the agreeable range detecting devours extra power amid its revealing stage. To this end, another plan elective for range detecting is to build up a geo-area officeholder database that keeps and intermittently refreshes the data of authorized ghostly inhabitance in some random geological area. Along these lines, unlicensed clients can promptly get to the accessible unused range by checking the occupant database with their current geo-areas. This is a vitality productive arrangement, however resulting in a debased ROC execution, since the range opening is resolved by utilizing a proliferation model of the essential flag, which neglects to reflect certifiable situations, for example, mountains, structures, and burrows. As a cure, we can join the helpful detecting and geo-area database, going for a decent tradeoff between the feasible execution and the vitality reserve funds.

8.7   Making Use of the Spectrum Network with the Use of Networking Cooperation

In this segment, we consider the advantages of cooperation in effectively utilizing the spectrum network recognized for vitality preservation in green cell systems. Figure 8.1 demonstrates cells arrangements comprising of the BS and UTs, which aids in assorted remote access interfaces, including both Wi-Fi and Bluetooth. Since the UTs are furnished with both Wi-Fi and Bluetooth, all are equipped for setting up an auxiliary system inside the cell system and aimed at optimizing the optional system for helping cell correspondences to accomplish vitality proficiency upgrades, where the optional system works inside the spectrum networks recognized. It is called attention to some institutionalization endeavors in creating Wi-Fi covering the vacant TV functions, which is apprehended as the IEEE 802.11af. This suggests forefront Wi-Fi and Bluetooth interfaces would be required to work in a progressively broad range involving frightful gatherings, as opposed to being restricted to the modern, logical and medicinal groups as it were. On the off chance that no range openings are distinguished, the optional system is deactivated and the UTs straightforwardly speak with the BS inside the cell range.

On the other hand, when a range opening is recognized with the guide of range detecting, distinctive remote systems might be conjured for participating with exemplary cell interchanges. This strategy may therefore be alluded to as psychological system participation, since the previously mentioned intellectual element was abused. While considering Wi-Fi for instance of the optional system, the UTs initially speak with a passage through Wi-Fi connects over the range openings recognized and after

Fig. 8.1 Cooperation and cognition with the aid of cellular wireless networking systems

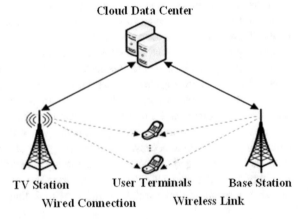

Cloud Data Center

TV Station   User Terminals   Base Station

Wired Connection   Wireless Link

that the AP trades information bundles with the BS through cell interfaces over the wireless range. Along these lines, the UT is in a position of communicating with the BS depending on AP used for enhancing the feasible vitality effectiveness of cell interchanges. As an option, we can likewise receive Bluetooth for helping cell correspondences connecting the UTs and the BS, in Fig. 8.1. Shows the loss of comprehensive articulation, let us consider the cell downlink transmissions from the BS to two or three UTs, decisively to U1 and U2. Initially BS was enabled to transmit its downlink data parcels to U2 and U1 over the cell range. Because of the communicating idea of the remote medium, U1 can catch the transmissions from the BS to U2 or U1.

At that point, when the range gap is distinguished, then U2 and U1 will trade their got flags through a two-way Bluetooth connect over the range gap identified. Along these lines, the work of Bluetooth over the range gaps gives spatial decent variety to cell interchanges and consequently lessens the general vitality utilization under an explicit target QoS necessity. Normally, a comparative system participation process might be connected using cell uplink deliverance from the UTs to the BS. To be expressed, given that a range opening is remembered, we at first allow the spatially scattered UTs to confer through a discretionary framework over the recognized range hole for exchanging their uplink packs. At the point when the UT gains each other's data packages, they can arrange for transmitting their bundles to the BS over the cell range. It may be seen that an auxiliary system is utilized for helping the cell interchanges using a sharp way, that a range gap has been distinguished.

This method is more complex than the conventional agreeable routine working in a homogeneous system condition. Unequivocally, in the conventional participation, an UT is required to transmit its information parcel to its helpful accomplice, which then advances they got information to the BS, the two transmissions happening over the equivalent cell range. This parts the cell range proficiency, because two symmetrical channels are necessitated for finishing the transmissions from a BS to the UT through an accomplice. On the other hand, insightful intellectual system participation enables an UT to transmit its information to its accomplice utilizing an optional system over the distinguished range gap, rather than utilizing the cell range.

The range detecting capacities are summoned for recognizing spectrum gaps, which are then assigned to the diverse system elements by the range. It might be seen that the two PHY-Macintosh sets share regular higher layers, specifically basic system and layers.

As an advantage, the diverse system get to crossing point can be facilitated and controlled with the guide of the upper-layer conventions. It merits referencing that contemporary cell gadgets ordinarily bolster various PHY-Macintosh conventions for collaborating with various remote access systems, including the use of Wi-Fi, 3G, Bluetooth, the LTE, and other diverse systems. By and by, the upper-layer convention the board is non-insignificant and ought to be considered for limiting the general system vitality dispersal. In this manner, it is exceptionally compelling to research the advancement of the collaboration plane by together thinking about the PHY, Macintosh together with the upper-layer conventions. In addition, it is critical to address the time scale and synchronization of issues during diverse system get to interfaces coordinate with one another.

8.7.1   Case Scenario: An Energy Cellular Communication That Uses Both Cooperation and Cognition

In this segment, a contextual analysis is conducted showing the achievable vitality sparing advantages of cognizance and collaboration systems in green wireless networks. Without loss of all-inclusive statement, as expected that in a network, the BS is transmitting its information bundles to a planned UT over the cell phantom band at a bearer recurrence of fc and inside the transmission capacity of BC. Therefore, the cell gadgets examine a television station utilizing spectrum detecting and recognize whether the checked station is inactive or not, where the transporter recurrence and the data transmission of the station are meant by Bt and ft, separately. On the off chance that the television slot is regarded to be inactive, it very well may be used for cell interchanges from multiple points of view, including the upgrade of the transmission unwavering quality or potentially the enhancement of the framework throughput. For instance, on account of an inert station, the BS rehashes the transmission of its information to the UT over the television slot by it's transmit control between the cell and television groups. Since these two frequencies are probably going to be adequately far separated to encounter autonomous blurring, they are equipped for giving a helpful recurrence decent variety gain, henceforth enhancing the unwavering quality as well as the vitality productivity, while keeping up the objective QoS prerequisite. Figure 8.2 demonstrates the adapted representation of joint cognizance and participation helped cell remote systems, where the station is used for helping cell transmissions of the BS to the UTs.

The BS and TVS are thought to be associated by a high-transfer speed optical spine or microwave connects, for instance through cloud server farm, which empowers provoke information trading between the BS and TVS. When the

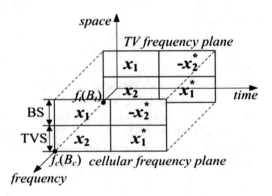

Fig. 8.2 Interlinked cooperation and cognition

Fig. 8.3 Untainted cooperation

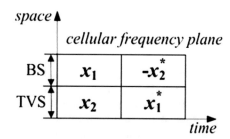

Fig. 8.4 Untainted cognition

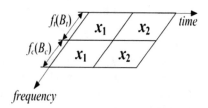

TVS got the BS' information bundles via the CDC, it can encourage the BS' transmissions to the UTs by using Altamonte's space-time code, as showed up in Fig. 8.3.

Even more unequivocally, the BS' data groups are utilized by x1 and x2, which are also known at the TVS utilizing the CDC. According to Alamouti's space-time coding, the BS and TVS, exclusively, transmit x1 and x2 to a normal UT meanwhile over a cell horrible band. At point in a space, the same time transmission of −x ∗2 and x ∗1 for the UT, as ∗ indicates the conjugate complex activity. This transmission requires participation between the TVS and BS, which works over the cell band, along these lines alluded to as the cell recurrence plane as appeared in Fig. 8.4. For reasonable examination, the aggregate power devoured at the BS and TVS is compelled to a settled dimension signified by P, particularly to the intensity of a solitary transmission by the BS as it were. In addition, an equivalent power-sharing is considered and transmit forces the TVS and BS as exhibited by $P/2$. Moreover, if the television station is observed to be inert, the BS and TVS may rehash the previously mentioned agreeable transmission over the inactive television station, called television recurrence plane.

To constrain this shared impedance to a middle of the road level, the discovery likelihood P_d and the false alert likelihood Pf ought to fulfill target esteems. The paper considers $P_f = 0.01$ and $P_d = 0.99$, except that it is generally expressed. In addition, let D_{ad} signify the likelihood that a TVS abandons a television slot and henceforth ends up accessible for cell interchanges. Where BS transmits the signal in force of PBS and at a data rate of R, the power gotten by the UT, which is demonstrated by P_R, which can be imparted as

$$P_R = P_{BS} \left(\frac{c}{4\pi \int cd_{BU}} \right)^2 G_{BS} G_{UT} |h_{BU}|2, \tag{8.1}$$

C is the speed of light, d_{BU} is the division of the UT from the BS, while f_c is the phone carrier repeat. G_{UT} is the get gathering mechanical assembly gain at the UT, G_{BS} is the transmit radio wire gain at the BS, and h_{BU} is the obscuring coefficient of the channel crossing from the BS to the UT. In this paper, the Rayleigh obscuring model is used for portraying the channel, hereafter h_{BU} with 2 that follows the exponential transport with a mean of $\sigma2BU$. It is normal that all beneficiaries experience Gaussian dispersed warm confusion with a zero mean and a change of $\sigma2n$ that is exhibited as $\sigma2n = \kappa TB$. Resultantly, κ addresses the Boltzmann steady while T is the system temperature in Kelvin and B addresses the redirect transmission limit in Hz.

Describing $N_0 = kT$, as the up leaval control thickness and considering a room temperature of $T = 290$ K, $N_0 = -174$ dBm/Hz. Using the data rate R in bits/s and the transmit control PBS in Watt, the imperativeness efficiency imparted in Bits-per-Joule is described to be appeared as follows:

$$\eta = \frac{R}{P_{BS}} \tag{8.2}$$

This is utilized for assessing the vitality cost as far as the quantity of bits conveyed for each Joule. It is seen from Eq. (8.3) that expanding the information rate R enhances the vitality proficiency η of the broadcast. Nevertheless, the blackout execution of cell transmission likewise corrupts, when a higher information rate is considered, since a blackout occasion happens more habitually after expanding the information rate. Consequently, there is a tradeoff between the vitality productivity and the blackout likelihood. As indicated by Shannon's coding hypothesis, a blackout occasion happens, if the channel limit falls beneath the information rate. As shown in the proposed joint comprehension and participation conspire, the TVS and the BS will rehash their helpful transmission to the UT over and inert television station recognized. Accordingly, utilizing the law of aggregate likelihood, the blackout likelihood of the joint comprehension and participation plan can be defined as

$$P_{out} = \Pr\{ C_{cell}^{coop} < R, \quad C_{TV}^{coop} < R, \quad H = Ho\}$$
$$+ \Pr\{C_{cell}^{coop} < R, \quad H = H1\}, \tag{8.3}$$

C_{cell}^{coop} conforms to the channel limit of the helpful transmission working over cell range band. C^{coop} television is the station limit of the agreeable transmission over the television range band, $H = Ho$ represents to the occasion of an inactive television station being distinguished, and $H^{\hat{}} = H1$ demonstrates that the station is esteemed to be involved by a TVS. Once more, attributable to the foundation commotion and impedance, the range detecting is inclined to missed location and to false caution

Table 8.2 Potential coordinative parameters utilized during numerical assessment

Noise PSD level	N_0	−174 dBm/Hz
Cellular carrier frequency	f_c	2100 MHz
Cellular spectrum bandwidth	B_c	5 MHz
TV carrier frequency	f_t	55.25 MHz
TV spectrum bandwidth	B_t	6 MHz
Data transmission rate	R	30 mbits/s
Antenna gains	G_{UT}, G_{BS}	0 dB, 5 dB
Transmission distances	d_{BU}, d_{TU}	1000 m
Average channel gains	$\sigma^2_{BU}, \sigma^2_{TU}$	1
Spectrum sensing performance	P_d, P_f	0.99, 0.01
Transmit power of TV station	P_{TVS}	45 kW

within the sight of a TVS. On the off chance that a missed discovery occasion occurs, the television clients and UTs will meddle with one another, i.e., the obstruction may incur a blackout occasion on the TV and UT inputs.

In reference to the notation comforts, assume the dTU and the PTVS communicate to the transmission intensity for the TVS which is a separation of UT from TVS solely. Additionally, with the remote of the channels of the TVS and UT is exhibited through the Rayleigh blurring, the normal blurring actuated lessening between the TVS and the UT is meant by σ^2_{TU}. Presently, we have reasonably described the connection between the vitality effectiveness and blackout likelihood of the proposed joint insight and participation conspire. It is brought up that we can comparably decide the vitality effectiveness and blackout likelihood of the unadulterated collaboration and unadulterated perception plans.

In what pursues, we present some numerical outcomes for describing the vitality proficiency utilizing Eqs. (8.1)–(8.3). Table 8.2 records the framework parameters utilized in our numerical assessment, where $f_c = 2100$ MHz while BC = 5 MHz contrasted to the 3GPP LTE frame. Consequently, in the North American region, about 2100 MHz of the LTE operates among diverse options for the channel distribution velocity. Within these frequencies, there exists about 12 TV centers for example center 2 to 13.

Thus work output executed by television slot two which comprises of a recurrence frequency of $f_t = 55.25$ MHz while the data conveyance capability $B_t = 6$ MHz for the numerical evaluation. The data rate provided is stated $R = 30$Mbits/s with the reception gadgets for the BS and the UT is thus exhibited by $G_{UT} = 0$ while G_{BS} and $D_b = 5$ dB. Then the transmission of the separations were spread over from the TVS and the BS to UT which is shown as $d_{BU} = 1000$ m while $d_{TU} = 1000$ m, individually the normal blurring gains are thought to be $\sigma^2_{BU} = \sigma^2_{TU} = 1$. Moreover, the prolific recognition probability and false alert probability of $P_d = 0.99$ and $P_f = 0.001$ in addition a transmit intensity of about $P_{TVS} = 45$ kW which are considered in our numerical assessment.

This unequivocally demonstrates that the vitality productivity of cell correspondences enhances as the blackout execution corrupts, inferring the nearness of a tradeoff between the vitality effectiveness and the blackout likelihood.

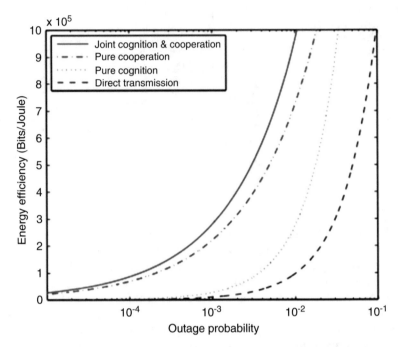

Fig. 8.5 Vitality productivity versus blackout likelihood of the customary direct transmission, unadulterated perception, unadulterated collaboration and in addition the proposed joint comprehension and participation plans with $D_{ad} = 0.8$

The vitality effectiveness of the joint perception and collaboration plot is reliably higher than that of the conventional direct transmission, unadulterated cognizance, and unadulterated participation as indicated in Fig. 8.5. At the end of the day, given a greatest middle of the road blackout likelihood with a unique count of the number of bits being conveyed, therefore the linked imminence together with the contribution with disseminates reduced vitality unlike the conventional express broadcast, with pure cognition and also with a complete cooperation of various approaches. Such a quantitative analysis exhibits the vitality merits of exploitation of the dual complimentary situation which are collaborative and perception within the cell systems.

Furthermore, both the unadulterated collaboration and the unadulterated insight accomplish higher vitality efficiencies than the immediate transmission; appearing both the participation and perception are useful as far as vitality utilization decrease. In perspective of the above contextual investigation, we may presume that the vitality dissemination of cell systems can be altogether lessened by depending on both discernment and collaboration methods.

8.8 Summary

The necessity to enhance the green wireless communication models continuously becomes more significant due to ubiquitous wireless networks. GWC will continue to provide more energy efficient wireless communication. Resultantly, minimal radiation from communication devices, which is a critical economic resolution for subscribers and service providers, will be evident. Various energy efficient initiatives such as the green manufacturing, BTS, green antennas, and the green handover solutions will be formulated and accorded to nature and human beings. This article analyzes the contextual evaluation of an achievable vitality sparing merits of cognitive radio in green wireless networks. In the case analysis, the BS is transferring its data sets in the UT over standardized cell phantom bands on a bearing recurrence of fc in BC's transmission capacity.

This chapter considers the merits of cooperation critically utilizing the spectrum networks in green wireless networks. Since the UTs are furnished with both Wi-Fi and Bluetooth, it implies that all are equipped for setting up a supplementary system inside the cell system and aims at optimizing the optional system for helping wireless networks to accomplish vitality proficiency upgrades, where the optional system works inside the spectrum networks recognized. It is called attention to some institutionalization endeavors in creating Wi-Fi covering the vacant TV functions, which is apprehended as the IEEE 802.11af. This suggests forefront Wi-Fi and Bluetooth interfaces would be required to work in a progressively broad spectrum.

References

1. Gür, G., Alagöz, F.: Green wireless communications via cognitive dimension: an overview. IEEE Netw. **25**(2), 50–56 (2011)
2. Lu, D., Huang, X., Zhang, W., Fan, J.: Interference-aware spectrum handover for cognitive radio networks. Wirel. Commun. Mobile Comput. **14**(11), 1099–1112 (2012)
3. Pesola, J., Pönkänen, S., Markopoulos, A.: Location-aided handover in heterogeneous wireless networks. Wirel. Pers. Commun. **30**(2–4), 195–205 (2004)
4. Gambini, J., Simeone, O., Bar-Ness, Y., Spagnolini, U., Yu, T.: Packet-wise vertical handover for unlicensed multi-standard spectrum access with cognitive radios. IEEE Trans. Wirel. Commun. **7**(12), 5172–5176 (2008)
5. Ma, B., Xie, X., Liao, X.: PSHO-HF-PM: an efficient proactive spectrum handover mechanism in cognitive radio networks. Wirel. Pers. Commun. **79**(3), 1679–1701 (2014)
6. Boujelben, M., Ben, S., Tabbane, S.: SON handover algorithm for Green LTE-A/5G HetNets. Wirel. Pers. Commun. **95**(4), 4561–4577 (2017)
7. Koh, S., Gohar, M.: Multicast handover agents for fast handover in wireless multicast networks. IEEE Commun. Lett. **14**(7), 676–678 (2010)
8. Mitran, P.: Interference reduction in cognitive networks via scheduling. IEEE Trans. Wirel. Commun. **8**(7), 3430–3434 (2009)
9. Yang, Z., Jiang, W., Li, G.: Resource allocation for green cognitive radios: energy efficiency maximization. Wirel. Commun. Mobile Comput. **8**(4), 1–16 (2018)
10. Durantini, A., Petracca, M.: Performance comparison of vertical handover strategies for psdr heterogeneous networks. IEEE Wirel. Commun. **15**(3), 54–59 (2008)

11. Li, Y., Cao, B., Wang, C.: Handover schemes in heterogeneous LTE networks: challenges and opportunities. IEEE Wirel. Commun. **23**(2), 112–117 (2016)
12. Jayaweera, S., Bkassiny, M., Avery, K.: Asymmetric cooperative communications based spectrum leasing via auctions in cognitive radio networks. IEEE Trans. Wirel. Commun. **10** (8), 2716–2724 (2011)
13. Oh, H., Ran, R.: Stochastic policy-based wireless energy harvesting in green cognitive radio network. EURASIP J. Wirel. Commun. Netw. (2015)
14. Ramamonjison, R., Haghnegahdar, A., Bhargava, V.: Joint optimization of clustering and cooperative beamforming in Green cognitive wireless networks. IEEE Trans. Wirel. Commun. **13**(2), 982–997 (2014)
15. Suganya, M., Anandakumar, H.: Handover based spectrum allocation in cognitive radio networks. In: 2013 International Conference on Green Computing, Communication and Conservation of Energy (ICGCE), Chennai, pp. 215–219 (2013)
16. Anandakumar, H., Umamaheswari, K.: Energy efficient network selection using 802.16g based GSM technology. J. Comput. Sci. **10**(5), 745–754 (2014)
17. Anandakumar, H., Umamaheswari, K.: Cooperative spectrum handovers in cognitive radio networks. In: EAI/Springer Innovations in Communication and Computing, pp. 47–63 (2018)
18. Anandakumar, H., Umamaheswari, K.: A bio-inspired swarm intelligence technique for social aware cognitive radio handovers. Comput. Electr. Eng. **71**, 925–937 (2018)

Chapter 9
Secure Distributed Spectrum Sensing in Cognitive Radio Networks

9.1 Introduction

The development of Cognitive Radio Networks (CRNs) and their applications has led to the congestion of the distributed spectrum in Primary Users (PUs) and Secondary Users (SUs). However, there are spectrum utilized by unlicensed users, which have been allotted specific privileges meant for the licensed users. Resultantly, this leads to the deprivation of the distributed spectrum's security in the CRNs and their applications. In order to eradicate such a problem, policy makers and supervisors should work on new spectrum management schemes. Distinctively, the American Federal Communication Commission (FCC) has been working on the issue in 3 different ways: spectrum sharing, spectrum reallocation, and spectrum management. In spectrum reallocation, there is reassignment of long-standing clients and regularity bandwidth to brand new wireless services, for instance mobile communication [1]. In leasing of the spectrum, there is relaxation of the commercial and technical constraints by the Federal Communication Commission based on the present spectrum accredit through allowing present licensees in utilizing their spectrum in a flexible manner which will be used in a variety of services or rather the leasing of spectrum to third parties.

On the other hand, in spectrum sharing, the FCC assigns the spectrum used for the unlicened or other linked services. While spectrum re-distribution and spectrum leases aim to improvising the accuracy of the utilization of spectrum based on the perception of the accredited administration of the spectrum, on the other hand, spectrum sharing focuses on improved efficiency of the spectrum utilization centered on the licensed perception of spectrum management, the objectivity of spectrum sharing is focused on regulating the unaccredited spectrum use. Particularly, spectrum sharing has drawn much curiosity from different regulators, producers, and other researchers. The Federal Communication Commission has considered bringing out a new spectrum sharing concept, in which the approved bands will be wide open

© Springer Nature Switzerland AG 2019
A. Haldorai, U. Kandaswamy, *Intelligent Spectrum Handovers in Cognitive Radio Networks*, EAI/Springer Innovations in Communication and Computing, https://doi.org/10.1007/978-3-030-15416-5_9

to the uncertified processes based on a non-interference basis within the accredited events. This is because some of the licensed bands for example the TV bands are underutilized; additionally the fallow segments within spectrum sharing of the licensed bands efficiently assuage the spectrum paucity issue.

The spectrum sharing hypothesis is known as the dynamic spectrum access (DSS); most licensed clients are known to be the incumbents or the primary clients, while the unlicensed clients who have a way in within the spectrum are thus known as secondary client. The expertise on Cognitive Radio Networks (CRNs) stages a vital role in the process of identifying DSS concepts. In order to attain highly elastic operational characters necessary for the DSS, the SDAs (Software defined radio) should be enacted within the CR as an alternative of the hardware centered applications specific integrated circuits (ASIC) strategies compared to the conventional radios. Consequently, the CR can easily learn based on the environment and intellectually amend its normal operational parameters centered on what was initially learnt [2]. Within the DSS, most CRs utilized by the SUs should be in a position of scanning specific spectrum range and smartly settling on a set of spectrum band in use for a particular transmission. Therefore such a process is known as spectrum sensing.

In an event of spectrum sensing, when secondary clients identify that it is contained within an incumbent shield zone 1 of a specific band, then it will abstain accessing such a band and search for a fallow band which is easily accessible. On the other hand, when there is no detection of the incumbents, then the secondary clients will link with other SUs in sharing the spectrum unused by the incumbents. Based from the exploitation event, secondary make use of the adhoc network architecture or the cellular networking architecture as indicated in Fig. 9.1. The composition of a WRAN cell is mainly a base station (BS)), coverage of WRAN cell, and the number of consumer premise equipments (CPEs) which ranges from 10 to 100 km.

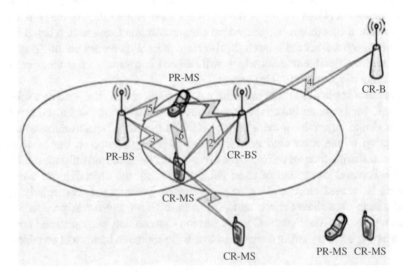

Fig. 9.1 Cognitive Radio Architecture: two links

The disparity in CR adhoc network consists of a reduced algorithmic energy level device which is featured with the CRs, which are freely associated with one another using multi-hop wireless links. Thus the institution of a particular set of wireless links is using the DSS. Though at times the two sorts of CR networking systems comprise of dissimilar networking architectures, in which the spectrum sensing is vital for the two, in which it corresponds to key technical obstacle which should be outwitted prior the prevalent operation of the CR networks is feasible.

Cognitive radio (CR) is a revolutionary technology that promises to alleviate the spectrum shortage problem and to bring about remarkable improvement in spectrum utilization. Spectrum sensing is one of the essential mechanisms of CRs and is an active area of research. Although the operational aspects of spectrum sensing are being studied actively, its security aspects have attracted very little attention. In this paper, we discuss security issues that may pose a serious threat to spectrum sensing.

Specifically, we focus on two security threats—incumbent emulation and spectrum sensing data falsification—that may wreak havoc in distributed spectrum sensing. We also discuss methods for countering these threats and the technical hurdles that must be overcome to implement such countermeasures.

9.2 Secure Distributed Spectrum Sensing Scheme

The application of distributed spectrum sensing schemes is centered on the SUs who are focused on distributed cooperative sensing whereby one user evaluates a collective estimate of sensing within a specific timeframe, denoted by "T." Relative to the timeframe scheduled for sensing, the success and development of distributed spectrum sensing security is assured by the SUs. Most secondary clients link up with common spectrum sensing outcomes together with their neighbors alongside the communication span and bringing up to date their personal ideals centered on the acquired values. In view of the fact that the distributed cooperative sensing may augment sensing precision, while minimizing sensing exactness, and high-priced technology, it is thus projected to improve the sensing execution process. Conversely, it is susceptible to inner assault threats. Thus the inner rivals can take control of various nodes when reporting to bogus detection outcomes in degrading the ultimate sensing decision, in which the execution of the cooperative sensing might diminish considerably.

Reputable systems have been vastly utilized in coping with false opinions based on a particular concept. The idea of reputation has been vastly utilized in social sciences, ecology, economics, and anthropology. On the other hand, vast section of literature has been set aside through the analysis of dissimilar reputable systems based on the computer networks [3]. Lately, depicted from the Dempster Shafer theorem with the capability of clearly managing and representing a node of hesitation, prejudiced logic and uncertainty based reputation methods have come up being eye-catching tool in managing trust affairs and have additionally drawn great concentration within the distributed CRNs. This segment, is proposition of the

DSCS, which is a reputable centered sensing scheme in which the distributed utilizing a subjective logical reputable framework in defense of inside harmful attacks on the secondary clients. A unique belief calculus is represented by a subjective logic which utilizes certain metrics known as opinions in expression of subjective standing. Due to the fact that it is compulsory in development if these mechanisms are used for detecting and management of the dangerous clients within the circulated CRNDs, therefore subjective reasoning is required with the capability of openly signifying and management of client's doubt which has recently come out to be an eye-catching device in maintaining trust relationship within the spread CRNs systems.

Considering this rationale on a logical aspect, the perspective shown by the tuple "$\omega x : y$" equals to $(bx : y, dx : y, ux : y, ax : y)$ which is signified by $bx : y, dx : y$ whereas U_{xy} controls the x node. This node signifies the disbelief while y shows the precise effect on the distributed spectrum sensing. Likewise, the a_{xy} base rate controls the x openness that assures the y evaluates the delays evident from the spectrum effect when an attack on the spectrum is evident [4]. The points of indecisions due to disbelief in the x node and y are assured by the 1.0 uncertainty indicating that nodes lack a fundamental basis of the sensing rationale. Upon utilization of any belief in making a decision, the belief/disbelief axis is thus prospected based on the outcome,$(\omega x : y)$, used in identifying harmful nodes.

Nonetheless, the connecting belief and disbelief, base rates changes over a particular timeframe and uncertainty assures the status evaluation of the SUs with the licensed clients. In that regard, the assessment of the users depends on the values and parameters of the PUs and the decomposed valuations of the previous base rates. With $\varpi directn, : y$ (bn, dn, un, an) being the direct view of client x to client y within time interval of t_n, it is kept as x's common reputable table and is calculated at x as shown below:

$$b_n \triangleq \sum_{c \in C} b(c, x : y)c = \frac{1}{|C|} \sum_{c \in C} \frac{N^s_{(n,x:y)c}}{N_{(n,x:y)c}}$$

$$d_n \sum_{c \in C} b_{(n,x:y)c} = \frac{1}{|C|} \sum_{c \in C} \frac{\left(1 - P^{\text{uncertain}}_c\right) N^f_{(n,x:y)c}}{N_{(n,x:y)c}}, \tag{9.1}$$

$$U_n \triangleq \sum_{c \in C} u_{(n,x:y)c} = \frac{1}{|C|} \sum_{c \in C} P^{\text{uncertain}}_c \cdot \frac{N^f_{(n,x:y)c}}{N_{(n,x:y)c}}$$

$$a_n = 0.5$$

In an event when $Nn, x : y$ $(Nn, x : y = Ns\, n, x : y + Nfn, x : y)$ is the overall number of the detection outcomes which x acquired from client y. Then $N^s_{n,:y}$ will be the number of accurate outcomes while $N^f_{n,x:y}$ will be the number of incorrect outcomes depicted from y, precisely. On the other hand, $an = 0.5$ symbolizes the case when x's increasing in supposing that y is about 50%. While p uncertain c symbolizes the extent of uncertainty in which y is reliable. Therefore, p uncertainty is calculated as shown below:

$$\frac{P \text{ uncertain } c = \beta_c.P_d.(1 - P_f)}{P_d = 1 - P_m},$$ (9.2)

In which an assumption is drawn since the channels are AWGN. β_c is thus symbolized as the change of likelihood of the channel conditions between 1–0 subsequent to y which has already sent various detection methods. P_d on the other hand is the sensing likelihood of secondary clients, P_f is thus the bogus alarm rate, while P_m will be the counterfeit negative rates. P_f and P_d are thus explained as shown below:

$$P_f = \Pr\{Y > \tau \,|H_o\} = 1 - \Gamma\left(\frac{L}{2}, \frac{\tau}{2\sigma_0^2}\right)$$
$$P_d = \Pr\{Y > \tau |H_1\} = 1 - \Gamma\left(\frac{L}{2}, \eta \frac{\sigma_0^2}{\sigma_1^2}\right),$$ (9.3)

In this case, H_o will be the nil hypotheses, showing that the detected channels will be inactive, whereas $H1$ is a substitute of the hypothesis, showing that the channel is full of activity. τ stands for the thresholds while Y is the result. In that case, the Signal Noise Ratio (SNR) is shown by $(\sigma\,21 = \sigma\,20)/\sigma\,20$. In an event L tests sampling figures, whereby n indicated satisfies equation $T\left(\frac{k}{2}, n\right) = 1 - pf$ which indicates the partially gamma application. In reference to the timeframe effect on spectrum decision, the valuation of time of licensed users within "t_i" at time "t_n" is indicated as:

$$b_n = b_i \times e^{-\left((b_i)^{-1}\Delta t\right)^{2k}}$$
$$d_n = d_i \times e^{-\left((d_i)^{-1}\Delta t\right)^{2k}}$$
$$u_n = 1 - b_n - d_n,$$
$$a_n = a_i \times e^{-1\left((a_i)^{-1}\Delta t\right)^{2k}},$$
$$\Delta t = t_n - t_i,$$ (9.4)

With k ($k \geq 1$) being the decomposition rate when determining the time transition rate of opinion. The interval amid ti and tn is shown as Δt ($\Delta t > 1$). When $\Delta t \to \infty$, $e - ((bi) - 1\Delta t)2k \to 0$ and thus $bn \to 0$. Therefore, it confirms that the recent proposition values comprise of minimum effect on the current opinion assessment since the time range amid them always rises [5]. The utmost dynamic decomposition time duration is thus shown below

$$\omega_{c,x:y}^{\text{final}} = \eta_1 \times (b_i, d_i, u_i, a_i) + \eta_2 \times (b_n, d_n, u_n, a_n)$$ (9.5)

$\eta1$ and $\eta2$ stand for the weight values with ($\eta1 + \eta2 = 1$, ($\eta1$, $\eta2 \in [0, 1]$)) being utilized in evaluating the extent at which the reputable assessment outcomes at time *tn* and time *ti* influence the dynamic time status. In an event when ($\varpi final\ n, x : y$) is low compared to the threshold, then the secondary clients will commence the recommendation repute and the last reputation assessment. Centered on the above stated dynamic reputable models and combination of the characteristics of the CRN, a reputable based safe cooperative sensing strategy (DSCS) is suggested. Within the DSCS, secondary clients associate with the detection outcomes while the final result of the cooperative group individuals assess the validity and channels for the increment of the sensing precision. Additionally, the DSCS is in a position of castigating the unreliable clients in order to minimize the effect of vague data within the network. Information about the DSCS is shown in Algorithm 1. Therefore, it is important to note that the *DB*local *X* is *x*'s known reputation table. On the other hand, the table size 1 Mb–10 Mb is thus determined by the number of cycles within the simulations, therefore the memory overhead should not be considered precisely focusing on the latest devices.

9.3   Cheat-Proof Strategy

The strategy assessed in this contribution is centered on the evaluation of perfect and complete distributed spectrum sensing information. Nonetheless, data like the channel gains and constraints is fundamentally safeguarded data in every single user, which indicates that no single guarantee is assured to indicate that the users are capable of revealing their safeguarded data to other users. Whenever cheating is significant, selfish users are able to cheat since it results to a significant payoff. According to the standardized cooperate protocols, which usually favor the users who have effective channel status, selfish users will also act to exaggerate their conditions to occupy more chances among the distributed spectrum. In that regard, embedding a cheat-proof strategy is a critical issue because of the disregarded and distorted data, which is capable of undermining the repeated games. Thus, there is necessity to formulate a dedicate scheme which is capable of testifying if the data given by a single user is the truth [6]. Though, this strategy and technique is difficult and complement during its implementation, mostly when viewed under a specific timeframe varying system. As for the standardized allocation protocol, there is flexibility in applying particular strategies to ensuring and inducing the truth among the users. In case MTT protocols are applied for the distributed spectrum sensing, it is fundamental to formulate a mechanism that will stimulate the players to be enforced to tell the truth by themselves. On the other hand, if the APF protocol is applied, the scheme centered on users' statistics is critical for preventing the users from falsification.

9.3.1 Mechanism Form Centered

Due to the fact that the MTT data sharing protocols assign the sources of spectrum to both the SUs and PUs with the upper and instant attained signaling power, all the players act in a manner that exaggerates their attained valuation. To finally evaluate the possibility of distortion in the exchanged data from the users is to enable all the users to tell the verity of their private information. The mechanism form strategy is applied to ensure that the users become honest when exchanging their data with others. Specifically, the users who expect and claim an extended valuation are requested to pay some tax, which normally increased according to the claimed valuation extension of the single user. However, if the users ask for a lower valuation, they will receive some compensation back, which is referred as a transfer in the Bayesian technique theorem. In an event when the transfer of a single player is negative, the user needs to pay other users; otherwise they should also get a similar treatment from the other users. Due to the fact that the respective users consider more on the money balance and the profits during the transfer and data transmission, the inclusive payoff is the achievement of the transmission inclusive of the transfer. This implies that after applying the function transfer, the distributed spectrum sensing games fundamentally become a unique game theory composed of original payoff that has been replaced by the inclusive payoff [7]. Through effective designation of the function transfers, the users are assigned a greater payoff after claiming their actual safeguarded valuations. In reference to the cooperative allocation protocols, the safeguarded data given by $g_i, g_2 \ldots gK$ needs to be exchanged between the users. Assuming that the timeframe slot denoted by $g_i, g_2 \ldots gK$ represents the representation of random variations $g_i, g_2 \ldots gK$ and the users observe their safeguarded data, then the users will be allowed to claim the "gi" value to other users. This valuation might be dissimilar from the other actualized figures denoted by "g"i, whereby all the active users claim their data concurrently. Because $g_i, g_2 \ldots gK$ is confirmed to be the standardized assumption, whereas, $g_2 \ldots gK$ is unconfirmed, the decision allotment and calculation transfer is centered on the claimed valuation compared to actualized figures.

According to the MTT spectrum sensing aspect, the users composed of the "di" index, $g_i, g_2 \ldots gK$ has the capabilities of accessing the channels which facilitates the information throughput in the actualized timeframe slot to be written in a compacted form as shown in the Eq. (9.6) below.

$$R_i\left(g_i d_i(g_i, g_2 \ldots gK)\right) = \left\{ \log_2\left(1 + \frac{g_i}{N_o}\right) \right\} \tag{9.6}$$

The ith user is the standardized cheat-proof strategy is shown according to Eq. (9.7).

$$t_i(g_i, g_2 \ldots gK) \triangleq \varnothing_i(g_i) - \frac{1}{K-1} \sum_{j=1, j \neq i}^{K} \varnothing_j(g_i), \qquad (9.7)$$

whereby

$$\varnothing_i(g_i) \triangleq E\left[\sum_{j=1, j \neq i}^{K} R_i\big(g_i d_i(g_i, g_2 \ldots gK)\big) \right] \qquad (9.8)$$

From the above expression, it is noted that the expectations of the users are considered over all the realizations denoted by $g_i, g_2 \ldots gK$ apart from g_i because the user is less skilled about the other users that are applicable in the progressing timeframe slot during the claim The expression denoted by $\varnothing_i(g_i)$ indicates the total summation of the dataset throughput expectations that every user has, which are denoted by "i" claiming a valuation of g "i". Instinctively, when a single user asks for a greater "gi" they have the capabilities of receiving a greater opportunity of accessing the distributed spectrum, whereas the remaining users will have a smaller share of the spectrum. Though, an increased payment to the users may possibly negate the successive profits from other spectrum access nodes, this is via the bragging of the spectrum framework attainment. Contrary to this information, when the $(g'i)$ compensation is claimed, the "I" user achieves a significant compensation which represents the overall costs which are less than what is being occupied in the distributed spectrum. In that regard, the equal portion of every user analysis reflects to their safeguarded dataset. Resultantly, a significant proof is given under a certain proportion, which is the standardized and planned mechanism. This is also an equilibrium whereby every user gives their private data and proof. So, for the purpose of proving the equilibrium level, it is necessary to be indicated in every "$i \in$" whenever all the inclusive players are available except when user "i" gives their safeguarded data without any forms of deformation. In this scenario, the most effective response from the users "i" is connected to the reported actual safeguarded data. Eliminating any forms of inclusivity, it is confirmed that the second users via user "K" report actual valuations.

9.4   The Incumbent Emulated Threats

At the point when an occupant is distinguished in a given band, all SUs abstain from getting to that band. In any case, when an auxiliary is distinguished, different SUs may share that equivalent band. At the end of the day, occupants have higher need than SUs in getting to spectrum assets. Within an incumbent emulation (IE) threat, a pernicious secondary user to other SUs through transmitting signals which imitate the feature of an incumbent. An outline of an IE assault is appeared in the lower left corner of Fig. 9.2. Because of the programmability of CRs, it is feasible for a foe to

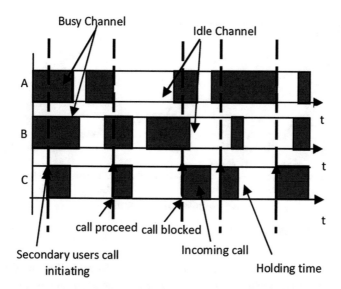

Fig. 9.2 Security threats within distributed spectrum sensing

alter the radio programming of a CR to change its emanation attributes (i.e., adjustment, recurrence, control). The potential effect of an IE assault relies upon the genuine SUs' capacities to recognize the assailant's signal from real occupant signals while directing spectrum detecting. Here we look at two existing spectrum detecting methods and clarify why they might be helpless against IE threats [8]. Vitality recognition is one of the most straightforward techniques for spectrum detecting. A vitality finder surmises the presence of an incumbent dependent on the deliberate signal vitality level.

Clearly, vitality recognition cannot recognize incumbent signs and auxiliary signs. An enhanced plan proposed in recommends the utilization of occasional calm periods. To encourage spectrum detecting amid a peaceful period, all SUs shun transmitting. At the point when calm periods are seen by all SUs, It recognizing occupants winds up direct that is, any terminal whose got signal vitality level is past a given limit can be viewed as an incumbent transmitter. In any case, such a recognition technique separates totally when vindictive SUs intentionally transmit amid calm periods. Signal highlight location is an elective method that utilizes either cyclostationary component recognition or coordinated channel identification to catch exceptional attributes of an incumbent signal. Be that as it may, depending exclusively on signal highlight discovery may not be adequate to dependably recognize an occupant's signal from those of an aggressor. For instance, in a CR spectrum where incumbents are television frameworks, an assailant may produce signals that imitate television signals. Then again, the assailant can replay television signals that were recently recorded. In either case, signal highlight discovery will dishonestly distinguish the aggressor's signal as that of an incumbent.

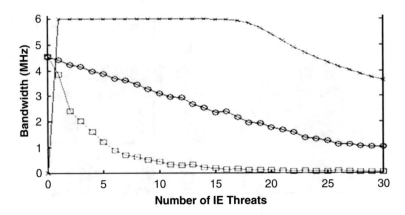

Fig. 9.3 Disruptive effects—IE attacks

A rival may have two unique thought processes in propelling IE threats. One inspiration is to pick up an called for favorable position in getting to spectrum in the spectrum sharing worldview of DSS. Since SUs will abstain from getting to a band if an occupant signal is recognized in the band, an assailant can seize and corner a neglected band on the off chance that it figures out how to trick others into trusting that it is an officeholder [9]. We allude to such an assault as a narrow minded IE assault. The second inspiration is to smother authentic SUs from getting to spectrum, in this manner causing refusal of administration. We allude to this assault as a vindictive IE assault. We did reproduction analyses to assess the problematic impacts of the two sorts of IE threats. Figure 9.3 demonstrates the recreation result.

We mimicked a 2000 m × 2000 m specially appointed CR organize containing 300 SUs, among which the quantity of IE aggressors changed from 0 to 30. There are 20 officeholder television slots, each with a data transfer capacity of 6 MHz and an obligation cycle of 0.2. While a self-centered IE aggressor planned to acquire at most one television band for its own utilization, a vindictive assailant propelled IE threats in all spectrum groups that were not utilized by occupants to amplify the problematic impact of the threats [10]. Our reproduction results show the adequacy of IE threats. Figure 9.3 demonstrates that the two sorts of IE threats can radically diminish the accessible data transfer capacity openings that each authentic auxiliary can recognize. As indicated by our outcomes, pernicious IE threats are progressively troublesome in diminishing the measure of accessible data transmission.

9.5   Protection Against the IE Effects

The way to shielding against IE threats is to devise a vigorous method for checking the credibility of an officeholder signal. One gullible methodology for confirming occupant transmitters is basically to implant a mark in an officeholder signal.

Another technique is to utilize a confirmation convention between an occupant transmitter and a verifier. These methodologies, be that as it may, are wrong in light of the fact that no adjustment to an occupant framework ought to be required to oblige artful range use by SUs. A potential arrangement exists when occupants are television frameworks. In a television framework, transmission towers are office-holder transmitters, where two properties can be utilized to recognize occupant signals from auxiliary signs. One distinctive property is the area of the transmitter. Since the area of a television tower is settled, on the off chance that the transmitter can be limited dependent on its signal, the area data can be utilized for confirmation.

In any case, an auxiliary found adequately near a television tower likewise would pass this area based check. At that point another distinctive property, signal control level, ought to be considered. The inclusion scope of a television tower normally changes from a few miles to many miles, and its transmitter yield control is regularly a huge number of watts. Conversely, SUs are hand-held CR gadgets that have a most extreme transmission yield control in a range from a couple of hundred meters to a couple of watts; this relates to a transmission scope of a couple of hundred meters [11]. On the off chance that an aggressor is in the region of a television tower, its signal control level would be fundamentally lower than that of the television signal. Along these lines, an officeholder signal transmitter can be checked utilizing a mix of the area of the television transmitter data and they got signal control level.

In this case the most difficult assignment is assessing or checking the area of the beginning of a flag. Since the DSS worldview endorses that no alteration to the officeholder framework ought to be required, the area estimation/check conspire must be non-intelligent that is, the area estimators/verifiers can't connect with the flag transmitter to gauge or confirm its area. Two strategies are exhibited to address the issue. The principal system is known as a separation proportion test (DRT), which utilizes got flag quality (RSS) estimations got from a couple of area verifiers (LVs) to confirm the area of the transmitter. A LV can be a devoted system gadget or an optional client with upgraded capacities to perform area check. Individual LV hubs frame a system and speak with one another. We expect that their information trade is anchored by a security convention. Since there is a solid relationship between the length of a remote connection and RSS, the RSS estimations at two LVs correspond with their separate separations to the area of the transmitter. The RSS esteem additionally relies upon parameters under the control of the transmitter, for example, the transmitted power esteem and the reception apparatus gain. In any case, when two LVs utilize indistinguishable radio beneficiaries and make synchro-nized estimations, it tends to be appeared under a reasonable radio engendering model; the proportion between their RSS estimations just relies upon the proportion between their separate separations to the area of the transmitter. One can figure out the normal proportion of the individual separations between each LV and the transmitter by utilizing the area data of the two LVs and the accepted position of the occupant transmitter. This proportion is contrasted and the proportion got from RSS estimations taken from each LV.

In the event that the normal esteem and the deliberate esteem are adequately close, the transmitter is viewed as an officeholder and passes the area confirmation;

else it comes up short the check. A noteworthy disadvantage of the DRT procedure is that its viability is impacted by the radio engendering model, which thusly is influenced by different ecological factors. Diverse proliferation situations may require the utilization of various parameters and may even require the utilization of entirely unexpected spread models. To address such issues, huge changes to the previously mentioned DRT procedure are required. The second method is known as a separation distinction test (DDT) [12]. This strategy utilizes the way that when a flag is transmitted from a solitary source to two LVs, a relative stage contrast can be seen when the flag achieves the two LVs because of their varying separations from the transmitter. For instance, when the occupant transmission arrange, in light of the fact that television signals include inserted inside them occasional synchronous images, two LVs can promptly quantify the relative stage contrast utilizing the beats or images.

The stage distinction can be converted into a period contrast that thusly can be converted into a separation contrast. One can compute the normal distinction of the particular separations between each LV and the transmitter by utilizing the area data of the two LVs and the accepted position of the occupant transmitter. This normal contrasted with the deliberate distinction with decide the credibility of the occupant signal. On the off chance that the two qualities are adequately shut, the transmitter is viewed as an occupant and passes the area confirmation; else it comes up short the check. In spite of the fact that DDT does not experience the bad effects of the disadvantages of a DRT, DDT requires tight synchronization among the LVs (on the request of many nanoseconds) that might be costly to execute. The past talk has been constrained to a situation where television frameworks are occupants. Moreover, future DSS applications might be reached out into other authorized groups, for example, those utilized by cell systems. These officeholders are portable and have low transmission control. Defining a compelling safeguard against IE effects that consider these kinds of occupants is an increasingly troublesome issue as shown in Fig. 9.4.

One conceivable issue is use the idea of radio condition delineates. REM is an incorporated database that comprises of far reaching multi-area data for a CR arrange, including the areas and exercises of radio gadgets. Given that such data is dependable and open to LVs, it is conceivable to confirm an officeholder transmitter by contrasting its watched area and exercises and those put away in the REM. More research is required to make such an answer reasonable.

9.5.1   Spectrum Data Detection Danger Falsification

Another security danger within the DSS is the broadcast of false spectrum detecting information by malignant SUs systems. An assailant may deliver false local spectrum detecting outcomes to an information authority, making the information gatherer settle on a wrong spectrum detecting choice. The term spectrum sensing data falsification (SSDF) is used in order to allude to such threats. Therefore, in order to

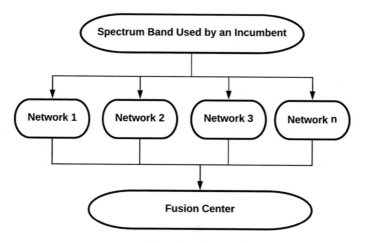

Fig. 9.4 Structuring the DSS as an equivalent fusion networking system

keep up a satisfactory dimension of exactness amidst the SSDF threats, information fusion system utilized within the DSS must be strong against deceitful local spectrum detecting outcomes detailed by the pernicious SUs. Despite the fact that a couple of data fusion systems for DSS have been proposed as of late, none tries to solve this issue. Accompanying such changes, we depict three information fusion systems that were proposed as of late for the DSS. This paper will depict every method briefly and examine its defenselessness to the SSDF threats. In order to facilitate this discussion, the DSS procedure is shown as a parallel fusion system, as exhibited in Fig. 9.4. Ni ($I = 0, 1, 2, \ldots, m$ with m being the quantity of detecting terminals) signifies a detecting terminal related with $N0$, which is both an information collector and a detecting terminal, y stands for the incumbent signal acquired at Ni, and ui is the nearby spectrum detecting outcome in which the Ni sends to $N0$ [13]. The result u is the last detecting outcome, which is a paired variable indicating the nearness of an incumbent signal, while zero means its nonappearance. In order to rearrange the discourse, the accompanying portrayal expects that spectrum detection is performed in a solitary band, with each ui being likewise binary.

The decision fusion wholly ties the majority of the gathered nearby spectrum detecting outcomes. Always thresholds limit of one and are no more than $(m + 1)$ should be indicated. In an event when the entire ui's is more prominent than or equivalent to the limit, at that point the last detecting outcome will be "occupied," that is, $u = 1$; generally the band is resolved to be "free," that is, $u = 0$. Since obstruction to officeholders ought to be limited, for the most part a preservationist technique is favored, which takes an edge estimation of one. For this situation, regardless of whether a band is free, insofar as there is one Ni that incorrectly reports $ui = 1$, the last outcome will be caught up with, causing a false alert. In the event that a SSDF aggressor abuses this and dependably reports one as its nearby range detecting result, at that point the last outcome dependably will be occupied. To forestall such a situation, one can expand the edge esteem. Be that as it may, expanding the limit esteem has the drawback of expanding the miss recognition likelihood. Besides, expanding the edge is insufficient in diminishing the false alert likelihood when there are numerous assailants.

Bayesian location requires the information of the earlier restrictive probabilities of $ui's$ when u is 0 or 1. It likewise requires the information of the earlier probabilities of u. Four cases must be viewed as $u = 0$ when a given band is free; $u = 0$ when the band is occupied; $u = 1$ when the band is free; and $u = 1$ when the band is occupied. Among the four cases, two choices are right, though the other two aren't right. The two right ones are designated with little expenses, and the wrong ones are related with extensive expenses. The miss discovery case is the least wanted situation and subsequently is doled out the biggest expense. The general expense is the entirety of the four costs weighted by the probabilities of the comparing cases. Bayesian location yields a last range detecting result that limits the general expense. At the point when a system is under SSDF threats, the estimations of the earlier contingent probabilities of the $ui's$ are not dependable [14]. Accordingly, Bayesian recognition is never again ideal regarding limiting the general expense. The Neyman-Pearson test does not depend on the learning of the earlier probabilities of u or any expense related with every choice case. It requires that either a most extreme adequate likelihood of false alert or a greatest worthy likelihood of miss identification be characterized.

Neyman-Pearson test assures that the other likelihood is limited, while the characterized likelihood is satisfactory. Similarly as with Bayesian identification, the Neyman-Pearson test additionally requires the learning of the earlier restrictive probability of the $ui's$ when u is 0 or 1. For a similar reason talked about already, the SSDF threats would undermine the ideal of the test and conceivably cause miss recognition or false alert occurrences. The recently referenced information fusion strategies share two properties in a manner that adds to their weakness of the SSDF threats. To start with, these methods handle all detecting terminals unpredictably, paying little mind to whether a detecting terminal is announcing valid or false detecting information. At the point when a SSDF assailant continually infuses false data, a perfect arrangement is to channel the information and just acknowledges contributions from solid detecting terminals. Secondly, the two systems can't ensure both limited false caution likelihood and limited miss recognition likelihood.

9.6   Protection Against the SSDF Threats

In order to counter the SSDF threats adequately, a two-level resistance is necessary. Within the main dimension, all neighborhood spectrum detection results must be confirmed by the information authority. The main reason for this safety precaution is to avert replay threats or false information infusion submitted by elements outside the CR network system. A second dimension of barrier is the arrangement of an information fusion strategy that is powerful against SSDF threats. As already examined, present information fusion plans are defenseless against the SSDF threats.

On the other hand, they can be enhanced in two different ways. One technique is through the utilization of successive probability ratio test (SPRT), comprising of information schemes that underpins a variable number of common spectrum detecting outcomes. The SPRT has the alluring property of ensuring both a limited false alert likelihood and limited miss recognition likelihood within a non-ill-disposed condition. Regardless of whether each detecting terminal has low range detecting exactness, the SPRT can give a certification by gathering increasingly common spectrum detecting outcomes [15]. Therefore this is preference over the methods talked about beforehand. The other method to expand power of the information fusion process is by bringing a notoriety based plan into the DSS procedure.

The structure of such a notoriety based plan can acquire knowledge from the current group of research on notoriety based secure directing plans for the impromptu systems. Instantly, a notable secure steering plan is proposed in employment of a two-module structure which is a "guard dog" module for notoriety support and a "path-rater" module for applying notoriety data to the routing systems. A comparative two-module structure can be utilized within the DSS, one for notoriety upkeep and the other for applying notoriety data to information fusion. In the initial module, a reputable rating is allotted to each detecting terminal dependent on the exactness of the nearby detecting report of the sensing terminal in respect to the last detecting choice of the information collector. Within the second module, the information collector applies the reputable rating to separate the "dependability" of the local spectrum detecting report got from each detecting terminal. Actually, there are numerous approaches to incorporate notoriety appraisals into a SPRT [16]. For instance, one strategy is to utilize the reputable rating as an example that can be added to the likelihood proportion of a detecting terminal. Such a plan guarantees, to the point that a detecting terminal with a reputable rating that is higher than that of different terminals assumes a more noteworthy role in settling on the last detecting choice. Therefore, the precision of the last detecting choice moves forward.

9.7 Summary

The distributed spectrum sensing aspect has experienced congestion from the users due to the enhancement evident in cognitive radio networks and applications. The wireless forms of application that operate with both the licensed and unlicensed frequency bands also include spectrum utilizing electro-magnetic spectrums assigned mainly for permitted players. Resultantly, this leads to the dispossession of the distributed spectrum for new and proposed cognitive radio applications. Although the performance of a dependable spectrum sensing approach is a tricky for the CRNs, various wireless channels and signals fading may potentially instill strength and security, which significantly lowers the proposed risks of the distributed spectrum models. The two fundamental forms of fading: multi-path and shadow reflected the signals which affects the dependability of the signals, strength and receiver location transitions. Moreover, the influence and effects posed on spectrum fading may potentially lead to the formation of hidden nodes, which is a significant problem in CRNs.

The security and hidden nodes issues in the spectrum sensing context of cognitive radio networks may be evaluated and illustrated as the instance whereby secondary users in CRNs are based in safeguarded locations where there are the incumbent users operating on the spectrums. However, the SUs in this context are not capable of detecting the subsistence of the present users. In that regard, the remedy for securing the distributed spectrum in the wireless networks is provided in the contribution in the forms of analyzing the cheat-proof strategy. Any subjective malicious nodes pose a threat on the spectrum channels, which are detected by the users as idle who need them for excusive utility. In an event when the users detect the spectrum channels to be too busy, they are not licensed to bear incentives for misreporting the channel data but report of true information in the fusion centers, which gives them the right to be non-trivial to context the distributed spectrum sense and any forms of spectrum falsification effect. For the purpose of neutralizing the attack of users misreporting the spectrum sensing status, the outstanding issues are formulated as a dual multi-standardized bandit procedure which applies random profitable match and the Bayesian methodology. In every circle, the spectrum fusion base chooses the subsets that are composed of the highly profitable nodes that can possibly be applied by their reports in the spectrum protocol whereas aiming at identifying the misinterpreted reports.

References

1. Lin, H., Hu, J., Huang, C., Xu, L., Wu, B.: Secure cooperative spectrum sensing and allocation in distributed cognitive radio networks. Int. J. Distribut. Sens. Netw. 1–12 (2015)
2. Anandakumar, H., Umamaheswari, K.: An efficient optimized handover in cognitive radio networks using cooperative spectrum sensing. Intell. Autom. Soft Comput. 1–8 (2017)
3. Anandakumar, H., Arulmurugan, R., Onn, C.C.: Computational intelligence and sustainable systems. In: EAI/Springer Innovations in Communication and Computing (2019)
4. Soto, J., Nogueira, M.: A framework for resilient and secure spectrum sensing on cognitive radio networks. Comput. Netw. 79, 313–322 (2015)
5. Taghavi, E., Abolhassani, B.: A two step secure spectrum sensing algorithm using fuzzy logic for cognitive radio networks. Int. J. Commun. Netw. Syst. Sci. 4(8), 507–513 (2011)
6. Wang, D., Ren, P., Du, Q., Sun, L., Wang, Y.: Reciprocally-benefited secure transmission for spectrum sensing-based cognitive radio sensor networks. Sensors. 16(12), 1998 (2016)
7. Anandakumar, H., Umamaheswari, K.: Energy efficient network selection using 802.16g based GSM technology. J. Comput. Sci. 10(5), 745–754 (2014)
8. Anandakumar, H., Umamaheswari, K.: Cooperative spectrum handovers in cognitive radio networks. In: EAI/Springer Innovations in Communication and Computing, pp. 47–63 (2018)
9. Yadav, R., Kumar, A., Singh, K.: Green power allocation for cognitive radio networks with spectrum sensing. IEEJ Trans. Electr. Electr. Eng. (2018)
10. Liu, X., Jia, M., Tan, X.: Threshold optimization of cooperative spectrum sensing in cognitive radio networks. Radio Sci. 48(1), 23–32 (2013)

11. Kim, M., Chung, M., Choo, H.: VeriEST: verification via primary user emulation signal-based test for secure distributed spectrum sensing in cognitive radio networks. Secur. Commun. Netw. **5**(7), 776–788 (2011)
12. Suganya, M., Anandakumar, H.: Handover based spectrum allocation in cognitive radio networks. In: 2013 International Conference on Green Computing, Communication and Conservation of Energy (ICGCE), Chennai, pp. 215–219 (2013)
13. Anandakumar, H., Umamaheswari, K.: A bio-inspired swarm intelligence technique for social aware cognitive radio handovers. Comput. Electr. Eng. **71**, 925–937 (2018)
14. Anandakumar, H., Umamaheswari, K.: Supervised machine learning techniques in cognitive radio networks during cooperative spectrum handovers. Clust. Comput. **20**(2), 1505–1515 (2017)
15. Haldorai, A., Ramu, A.: Cognitive social mining applications in data analytics and forensics. In: Advances in Social Networking and Online Communities (2019)
16. Jones, R., Veendorp, E.: Cooperative moves in a non-cooperative game. Global Business Econ. Rev. **7**(1), 25 (2005)

Chapter 10
Applications and Services of Intelligent Spectrum Handover

10.1 Introduction

The spectrum forms a collection of different forms of electro-magnetic radiations composed of feasible frequencies and wavelengths applicable in wireless communication. The intelligent spectrum frequencies define the natural asset, which does not deplete upon utilization by users. However, these resources render useless when not utilized effectively, hence implying that the handover of the spectrum is critical since it assures the free interference of operations in every radio applications. Approximately all countries depend on a similar reserve and electro-magnetic spectrum on the basis of individual rights defining unlimited utility. To facilitate the model of hardware compatibility and setting, standardization, setting and interference of free-wireless communication, it is critical to allot the spectrum based on similar spectrum bands [1]. The global communication union allots the frequencies of the spectrum that are fundamental for utility by other nations. Management and development of the intelligent handovers is responsible for the enhancement of the spectrum coordination, which includes the permitting and handling of the wants of various telecommunication users in each nation. The control of spectrum control is responsible for the operation of various wireless communication stations, which makes it necessary to launch an advisory group for the allocation of the spectrum frequency since it is a critical element in enhancing wireless communication. This advisory committee is tasked with the role to recommend fundamental allotment of frequencies and issues, which further propose the need to formulate plans for spectrum allotment and proposals for different problem mitigating factors connected to the intelligent spectrum handover and unlicensed users. Systems in wireless communication are formed on the broadcast of electro-magnetic waves characterized

© Springer Nature Switzerland AG 2019   193
A. Haldorai, U. Kandaswamy, *Intelligent Spectrum Handovers in Cognitive Radio Networks*, EAI/Springer Innovations in Communication and Computing,
https://doi.org/10.1007/978-3-030-15416-5_10

of frequencies ranged about 3 Hz and composed of different features of spectrum propagation. The frequencies are each defined on a certain telecommunication application such as wireless communication, cognitive radio navigation, and mobile communication.

10.2 Spectrum Scarcity

Considering the number of users that are growing at an exponential speed in reflection of the spectrum scarcity in the telecommunication market, there is necessity for an addition of more spectrum resources to control the rapid rise. The increasing rates of data are also necessary to enhance the quality of services for various services of the intelligent spectrum. Moreover, there are numerous information applications that utilize more bandwidth characterized and caused with an increasing amount of users in mobile communication dependent of the services and applications. Resultantly, it is fundamental to speed up the continued upgrading and expansion of the nation's intelligent spectrum handover and broadband connections [2]. The transforming speed of the global utility of mobile application, technological enhancement, and economic lifestyle necessitates the utility of more sophisticated services and application, which is speculated to enhance the driving demand on the spectrum and mobile services. The developing demands for enhancing the various applications of intelligent spectrum handover are the backbone for the large spectrum necessities; therefore, different wireless communication service providers have mainly been considered as longer heads for one another based on various spectra; however, more spectra is still unutilized in different spectrum bands which are applied in intelligent communication.

According to an analysis conducted by the Federal Communication Commission (FCC), there is the need for channels to be composed of minimal utilization of the spectrum. Intelligent spectrum is a critical asset, which implies that its underutilization of a greater extent on the band is never affordable. However, the ideology that supports the shortage of the intelligent spectrum is rapidly expanding due to present set policies and procedures define the permission of using the spectrum. In that regard, the increasing demand for spectra might be nearly impossible to meet when the alternating regulation scheme is still not recognized since what is needed is the critical strategy that permits the unpermitted individuals to operate in permitted spectra while accommodating existing individuals at a greater extent. The theoretical framework of abundant white-spacing accessibility and enhancing the intelligent spectrum handover causes the authorities such as the FCC to propose a rule behind the opening of underused spectrum for unpermitted utility. Therefore, the necessity for rule-making originates from the analysis of various forms of devices which operate in unpermitted spectrum eliminating any interference in permitted individuals, known as the primary users.

The unpermitted transferrable lower-powered devices which might conform to the WLAN card in PC and the constant extreme-power devices are applicable when delivering business services such as the broadband access to the internet. Wireless communication is thus introduced as a remedy which mitigates their status, whereby cognitive radio networks become a node that is capable of sensing its environment that allocates various intelligent spectrum chances. These opportunities utilize the spectrum hole to transfer the data and time, thus enhancing the utility of spectrum, whereby cognitive radio accepts unpermitted telecommunication users to get access to permitted spectrum from holder who are legit based on a certain level of negotiation. This action is fundamental since it presents a platform for cognitive radio, whereby a node or network enhances the reception and transmission of parameters to communicate effectively while eliminating a certain degree of interference with either unlicensed or licensed users [3]. The capabilities of a particular device to study its ecological perspective and enhance its means of adaption and performance development permit the changing from an oversight and manual process to an automatic and device-oriented process. Its capabilities also have the possibilities of allowing more sensitive utility of the band by means of minimizing the accessibility of the spectrum and its entry barrier of different services and devices.

10.3   Services Analysis of Intelligent Spectrum Handover

10.3.1   Features and Capacities

In order to increase the utility of resources, the upcoming generation of networks must take advantage of the sophisticated devices based on intelligent spectrum since this is fundamental for modeling the radio demographic area of telecommunication users, ecological aspect of users, and the entire networking framework. Centered on the setup and monitoring of various parameters, intelligent spectrum may fundamentally adapt various effective bands of frequency, interfaces, and protocols. In analyzing the service features of spectrum in adapting the broadcast parameter in the transforming ecological aspect via cognitive cycle, there are six fundamental cognition stages. These are:

- Observation: This means the identification of data in the operative ecological dimension via signaling and sensing techniques.
- Orientation: In this stage, the users need to analyze the relevant data for determining their relevance and significance.
- Planning: Centered on the evaluation, cognitive radio is capable of enhancing and determining different alternatives and options leading to the optimization of resources.
- Deciding: This level allows the choosing of one alternative, which is critical for evaluating the best options and present actions.

- Acting: Intelligent spectrum applies different decisions, which are critical for the optimization of resources. As a result, these transitions are launched on the profiles of the CR.
- Learning: In the entire education process, the cognitive radio applies its results during the decision process and observation in enhancing various operations and formulating unique alternatives and model status.

Services of intelligent spectrum handover enhance the efficiency of the spectrum in telecommunication system. The wireless nodes and networks are the basis of reception and transformation of the parameter to enhance their communication in eliminating any forms of interference with both the permitted and unpermitted users. Thus, this necessitates the monitoring of various factors based on the inner and exterior cognitive radio ecological dimension like the behavior of the users, CR frequency, and networking status [4, 5]. Cognitive radio networks may fundamentally depend on spectrum width to a greater extent, enhance the detection of spectrum holes and utilize holes critical for communication when necessary, but it does not have to affect the primary users. The whole context of intelligent spectrum in this instance is referred as the secondary users, whereby the interference of air for cognitive radio is centered on four fundamental states: sensing, management, mobility, and sharing. The intelligent spectrum is a critical and scarce resource, which is authorized and allocated for utility by the relevant authorities of the state. Normally, a single band of frequency is applicable for a greater extent by a single system in telecommunication, thus forcing various communication systems to apply various spectrum bands, which eliminates any forms of interference over the bands.

Nevertheless, the resources in intelligent spectrum are progressively scarce due to various spectrum-centered devices and service, which are developing markedly due to resultant response of wireless communication. However, based on the state management of the band strategy, the handover of intelligent spectrum considers the initial actions that will be considered for long-term utility by permitted system capacities though resources are limited and invariable. In various bands of spectrum, various services appear in greater volumes since many of mobile terminals apply these networking resources and can utilize various data and voice services on spectrum handover to the desired systems. In reference to the expanding amount of terminal, the CR systems are subjected to intense overloading that will potentially lead to the downgrading of network performance and lower the satisfaction of both primary and secondary users. The level of downgrade can be controlled by the current networking operators. However, there are particular technological initiatives like loading balance, admission controls, and band happing that are launching purposely for eliminating the possibility for overloading [6]. The policy controlling the basis of intelligent spectrum performance due to fact that when the capacity of the systems is feature an invariable, the users which need to be carried by a particular network may pose a reaction of a higher dimension. Apart from that, the resource of the spectrum is diminished due to some frequency spectrum rendering unutilized for a particular timeframe. The traffic that originates of telecommunication composes of the radio framework transitions defining a place or time. However, the band defining the present cognitive radio framework handover is consideration of the increasing traffic, which causes the shortage of spectrum assets during off-peak time.

An ideology behind spectrum handover is centered on cognitive radio, which was initially introduced by Mitola. This technological aspect facilitates the application of band resources at any place and time since it broadly deemed as a complete solution when lowering the application of wireless spectrum, which applies the problems arising between the extending services of wireless communication and the spectrum scarcity. Cognitive radio initiative will transmit the work modes from common configurations to an intelligent spectrum control in the telecommunication framework. Resultantly, the wireless system will be made a critical section of the global market economy, including the transformation of modes of spectrum management of resultant wireless networks during band management of rules and regulation [7]. Currently, the initiative is creating unique challenges that need to be tackled by manufacturers of devices, users and operators of networks. In that regard, to enhance effective utility of the intelligent spectrum resources, the handover of the spectrum is fundamental to limit the amount of wasted spectrum. The problems presented as a result of technological advancement are composed of more technologies necessitated for the performance of intelligent spectrum handover like the spectrum sensing development, interferences suppressions, and control of power technology. The economic problems may define the manner in which the economy incomes that are related to intelligent spectrum over band owner are managed. This involves the handover of the spectrum, its mechanism and algorithms like the public-sale algorithm and band lease activity.

Considering this trend of development, the networking devices and terminals need to be reconfigured, which further implies that they need to support the work conversion frequencies to sophisticated parameters to enhance their capability of utilizing more wireless communication resources stimulated by the handover of intelligent spectrum. Moreover, the application of reconfiguration technologies is critical due to its flexibility and effective for enhancing the performance of intelligent and rapid spectrum frequencies handover in a particular base station. Currently, the traditional reconfiguring wireless terminal coexists with a reconfigured mobile terminal for a greater timeframe, whereas the initiatives develop despite the fact that the terminals hold the reconfiguration of the terminals [8]. Resultantly, the forms of services are necessitated when setting various desires of a structured band frequency during the handover. Considering the technological aspect illustrated above, the systems can categorically obtain a minimal leasing spectrum via the interlinked network negotiation model that detects the band hole with the application of the cognitive radio technology. Moreover, this technological aspect also aids in accessing the existence of more spectrums whenever an open spot avails itself, which further replenishes the capacity of the system.

The existence of the intelligent spectrum handover strategy is applicable to intelligently allocate various frequencies when interference has not been detected and when the issues of spectrum frequency and handover have not been addressed. With the application of the present technological aspect, the management and utility of the band spectrum are evaluated long-lasting and identifies spectrum data resources during the handover of the spectrum. The rationale of this evaluation is due to uniquely allocated resources which is heterogeneous and make-shift, mostly for the acquisition of band resource data. The information is retrieved by means of spectrum sensing, which is unstable and in time-constraint; and also uncertain and incomplete. Due to the fact that the handover of the spectrum will reveal some possible delays and lead to transition of the structure of the network, the quality of service and application needs to be assured when the process of handover is in progress. In reference to the present system, many terminals like software and Hardware, lack the supporting illustration of spectrum frequency bands and this resources are render ineffective and useless.

Even when the software and hardware including the bandwidths and speeds conform to all protocols of spectrum handover, there are services unsuitable due to the make-shift of the frequencies and their stability which renders it ineffective for when considered using the quality of service standards. When the handover is strained in this manner, any unrequited issues and calling drops will potentially deliver the results. One of the issues to mitigate is the application of uniquely gotten spectrum resources which permits more users to boost and access the throughput of the system. Considering this instance of one cell in the telecommunication system, the capacity of these systems will be finished by the expanding amount of users, which implies that the system overload will be incompatible with the progressive sessions. As control of this issues due to scarcity in the system capability and cognition is centered intelligent spectrum handover. This fundamental as a technique which is applicable for sensing and intelligent handover or reconfiguration of initiatives considering characteristics of limited frequencies and features of service. Under this strategy, the success degree of the intelligent spectrum handover is speculated to develop, whereas the blockage status is expected to be lowered, thus boosting the performance of the wireless communication systems according to evaluation of quality of service.

10.3.2   Cognitive-Centered Spectrum Handovers

When the systems are available with frequencies handover range utility, stations and terminals they instigate the frequency function reconfiguration and detected band resources, which are sufficient. In this instance of individual cells, whereby the system loadings exceed a particular threshold, the intelligent spectrum system retrieves a limited timeframe band resource which eases the loading pressure. This replenishes the capacity of the system including the realization of the intelligent allotment of hand resource via the inter-networking resource negotiation and leasing or band holes based on terminals. Though, the resource frequency attained by a means of intelligent spectrum in the making shifts and analysis of data which is unstable, uncertain, and incomplete. Moreover, when the resources are attained via spectrum sense, the intelligent spectrum permitted for such resources may take advantage of the resources at any moment which implies that the system that accesses it makes shift of the resources. Moreover, the system is also capable of getting ready for the handover into a relative unused frequency at any moment. To further explain this, a collection of new issues come up due to the utility of unique handover resources.

For instance, when the terminal supports the spectrum frequencies, even if the forms of services match the feature frequencies, the frequencies are carried by the spectrum frequency, including reallocation flow from the effectiveness of time. In order to mitigate this issues and any resultant call drop following the handover, the terminal is chosen to facilitate the handover of the spectrum, which stimulates the necessity for some issues to be considered. The critical issues put under consideration include the configuration of terminals, service bands, and transforming speed irrespective of whether the services are transferred through the terminals or not. This also features the terminals relevant for the targeted frequencies [9]. Due to the fact that non-reconfigured terminals do not utilize the unique spectrum assets, the efficiency of the spectrum cannot be enhanced based on this instance. Moreover, it is fundamental for unique resources to be persevered when waiting for new-fangled sessions. Other than reconfiguring various wireless terminals together with live sessions in unique spectrum bands, it is fundamental to adapt and handover unique resources of the spectrum. Thus, the success rates may be enhanced making the original permitted spectrum frequency to accept common non-configured terminals, which causes an extensive throughput and organization capacity.

The services of intelligent spectrum handover are classified into four main forms of service according to the current wireless systems. These classes are:

- Interactive services
- Conversational services
- Streaming services
- The back services

The mentioned forms of services are composed of different necessities based on delays during the handover process. For instance, the streaming and the conversational services are characterized for their sensitivity during delays, whereas the back and interactive services are less sensitive when it comes to detecting delays. These services have parameter: bandwidth, delay requirements, and the time service estimation relevant for defining the features. Various terminals of users are characterized with various forms of services composed of different delay necessities, which reflects to the forms of services defining the stability of the spectrum. Considering various delays necessities, the services mentioned above are adapted and categorized according to allotted spectrum resources. The minimal the delay, the more firm the band is standardized to be, which further reflects on the suitability of the spectrum, composed of all the information based on very effective state. For instance, the streaming and conversational service which are delay sensitive are handed and selected over considering the preference of limited spectrum characterized by defined data. The services which are delay sensitive, e.g., the back and interactive services, are chosen and handed to the spectrum with a limited timeframe data, which is stable. As for various forms of services which do not necessitate stable frequencies, the duration of service is an estimate and calculated in sequence form [13].

10.4   Intelligent Spectrum Sensing

Based on the perspective of cognitive radio, the sensing of intelligent spectrum defines the discovery of band holes using spectrum detection methodologies, interference-based detection, match filter, and the cooperative sensing. The capabilities of cognitive radio networks to exist in various present networks may be predicted from its possibilities to sense the availability of surrounding networks. It is not also necessary for intelligent [14] spectrum to sense the holes since progressive spectrum monitoring is fundamental for search the primary user. The sensing dimension, timeframe, and accuracy are the features fundamental for considering the sensing of resultant issues during the handover of the spectrum. These interlinked issues include falsification during alarming which arises when sensing the availability of primary users in the area, hence making it possible for intelligent spectrum to sense users even when there are no primary users. The second connected issue is the missed alarm which arises when the primary user is available in the vicinity and is incapable of sensing the availability of the users [15].

10.4.1   Supervision of Spectrum

The supervision of spectrum basically entails being access to excellent availed spectrum which are in a position of attaining every client's communication necessities. The main utility includes evaluation of the spectrum and choosing a particular set of band based on each client's necessities. Therefore the CR systems should be in a position of adjusting to the brand allotted resources through brilliant utilization of the nearby resources or rather the ones that have been identified. In order to initiate accurate optimization and stimulation of the events, the CR should clearly organize, represent, and harbor the acquired knowledge and skills effectively. Consequently, certain type operating and transmission parameters should be evaluated keenly in order to attain maximum combination of such parameters which in turn get tuned in order to sustain the QoS. Various optimization methods have been utilized such as artificial intelligence and other algorithmic soft methods [16].

10.4.2 Multimedia Services

This scenario has been witnessed on an international auction level based on the vestige of the radio spectrum. Radio networks have been allocated to dissimilar networking sites, thus it is cumbersome based on the trending wireless networks in accessing the services due to the presence of a rigid spectrum rules and regulations and a high opportunity cost. Due to the inadequacy of the existing spectrum management, it thus leads to the shortage of artificial spectrums. Researches carried out worldwide have exhibited that a huge amount of radio spectrum is always underutilized. The DSA (dynamic spectrum access) together with the spectrum reframing systems are the best solutions in handling the issue of spectrum scarcity within the systems.

Within the DSA, unqualified clients opportunistically utilize the available licensed spectrums using the cognitive radios. The main essential component of the DSA is the cognitive radio. On the other hand, the theory behind cognitive radio is that secondary users (SUs) for instance the unqualified clients can be able to point out the free licensed spectrums for the qualified clients, e.g., the primary user (PU), and thus utilizes it despite that detrimental influence within the PU system. Additionally, a secondary application of the licensed spectrums makes use of effective utilization of the spectrums. Therefore, spectrum reframing entails an upturn of several spectrums based from the current users with the aim of re-assigning, either for brand new clients or rather for the emanation of fresh spectrally effective technologies.

10.4.3 Mobility of Spectrum

The handover process of spectrum basically refers to the amendment of the operational frequency or bands. Therefore, mobility happens when the CR diverts its frequency bands during the sensing of the PU signal waves. Therefore, the CR should be able to control a different frequency, through the preservation of seamless communication requirements throughout the transition in order to acquire an efficient spectrum as in Table 10.1.

The main purpose of this function is attaining an absolute QoS possibly [10]. SNR latency, data rate, and throughput are inclusive characteristics which are essential parameters during the process of choosing the right handover used in acquiring a seamless connectivity.

Table 10.1   Service parameters values for the IEEE 802.22

S. No.	Parameter	Value
1.	Spectral efficiency	0.5–5 bit/s/Hz
2.	Average spectral efficiency	3 bits/s/Hz
3.	Throughput	(a) Downstream—1.5 Mbps per CPE (b) Upstream—384 Kbps
4.	Coverage	100 km
5.	Operational frequency range	41–910 MHz
6.	Channel bandwidths	6, 7, and 8 MHz
7.	Threshold for vacating channels	−116 dBm over a 6 MHz channel (digital TV) −94 dBm at the peak of NTSC
8.	Wireless microphone	−107 dBm in 200 KHz bandwidth
9.	Services	Voice, data, audio and video

10.5   Spectrum Sharing

After realizing that the CR conveys the frequency, it on the other hand communicates with the receiver about the selected band so that a unique communication channel is introduced. Moreover, a reasonable spectrum scheduling protocol should be provided. This can be considered to be the same as the generic MAC issues within the available systems.

10.5.1   Consistency and Execution

Due to popularity within the CR networks, many business enterprises have emerged enforcing the consistency within the CRs steadfast functionality and its worldwide implementation. Various standards that are being structured along with its primary features are stated below:

10.5.2   The IEEE 802.22

During the year 2004, there was formation of the IEEE committee. This was the latest IEEE committee whose main objective was focused on the structuring of the WRAN (wireless regional area network) making use of the CR and thus coming up with basic requirements for the air interface. Such networks through the use of white gaps within the allotted television frequency spectrum. The outcomes based on the studies of the use of the TV bands were ease in the detection of the incumbent, and the dis-involvement within the life decisive applications initiated by the television bands use within the WRAN networks. Consequently, the WRAN may have an edge more than other types of networks based on the area covered since the main target is providing a coverage that is about 100 km alongside 15 km which is a maximum coverage provided by the great WAN networking systems [11]. There is certain proposition that the WRAN networks ought to have a greater coverage within the wireless networking systems lately. The main reason behind this is a greater power and more conducive propagation feature of the TV bands. In an event when the power supply is the main problem, then the coverage of the BS will relatively rise to about 100 km. Some vital parameters of such standardized systems are clearly exhibited in Table 10.1 in this paper.

10.5.3 The IEEE 1900

In the year 2005, there was formation of the IEEE P1900 paradigm committee which was initially founded through a mutual agreement of both the ComSoc (communication society) and the EMC (electromagnetic compatibility society). Therefore the main objective of the committee was the structuring of basic standards sued for the next generational radios and the improvised spectrum management. The IEEE 1900 comprises of main 3 groups which are standardized; they include:

- 1900.1: the description of terms linked to the CR networks
- 1900.2: accessing and the evaluation of the CR network operations
- 1900.3: different techniques used during the licensing of the software modules
- 1900.4: the network device supply decision for architectural structures used for CR optimization
- 1900. A: authorization of certificates for the CR networks

Currently the final version of the IEEE has been released. Afterwards an IEEE standard synchronizing committee was set up in 2007 (SCC 41). Therefore the committee was made up of certain IEEE 1900 operational groups which are in turn supported by both the EMC and the ComSoc.

10.5.4 Other Enterprises

Also other types of firms and groups, functioning mutually with an aim of standardizing and advancing the CRNs, include the DARPA (defense advanced research project agency), the ITU (international telecommunication unions, the WWRF (wireless world research forum), and many more. Fascinated readers also refer this to being a general outline of events of such standards as shown in [15].

10.6   Intelligent Spectrum Application Sectors

Since wireless gadgets such as the smart phones have covered a huge segment in our daily lives, therefore there is great need for connection to any individual at any time and in any place. As such it is an exciting challenge within the wireless research communication systems since there is need for the analysis of brand new methods used for taking advantage of the restricted radio spectrum bands. A set spectrum assignment scheme utilized by the present wireless networking systems is actually ineffective. Thus a field work carried out by the New York city department found out that a greater number of spectrum occupancy is about 13.1% within a 30 to about 3 MHz band. Consequently research has exhibited that about 22% of the allotted spectrums is used in the urbanized sectors unlike in the rural areas which is used at an extent of about 3%.

Thus the FCC (federal communication commission) approximates that the utility level for the present spectrum ranges from about 15% to about 85% which is a higher variance of geographical location and the time duration. The CR networks exhibit a high assurance for the usage of such deficit through the application of dynamic spectrum problems. The CR network is a new theory set up within the wireless communication system which permits spectrum deficiency which should be effectively used in an opportunistic way. Except that the traffic is unremitting and consistent, then an assigned spectrum band usually is underutilized or rather not used, thus resulting to the ravaging of the radio resources. Thus the unused spectrum bands are therefore referred to as the spectrum gaps (white spaces). Spectrum holes are thus subdivided into two major sections as illustrated by Fig. 10.1 precisely.

Spectrum gap is a segment comprising underutilized spectrums which are not assigned also Certain set of assigned spectrum for the time period. Presuming that the utmost throughput of the band will be 20Mps, it thus represented a blue dash line. As exhibited in the diagram in the paper, there is maximum utilization of the spectrum band which is rarely fulfilled, which lets go of huge white gaps within.

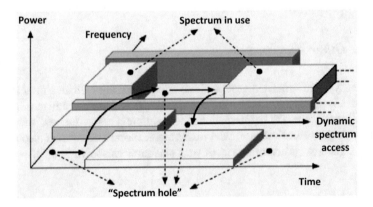

Fig. 10.1 Illustration of spectrum holes

The underutilized white gaps is termed type A spectrum holes. Taking advantage of this type of spectrum is in need of complicated spectrum distribution. There is full utilization of the spectrum band as exhibited, although there comprise of pauses within the transmissions, thus generating a vast number of white gaps. Therefore this is known as type B spectrum gaps. Such a spectrum is much easier for detecting and using it.

In real both type A and type B usually relate based on the assigned spectrum band whereas type A happens often compare to type B. Within a predetermined spectrum assignment, such spectrum gaps are in a position of filling up to about 85% of the allocated band. Thus clear analysis of the spectrum gaps enthused the prologue of the CR networks. There are efforts instilled by the graph hypothesis in the minimization of the spectrum assignment to being a variant of sensible coloring graph problem. In this technique different channels are symbolized using dissimilar colors while the SUs are thus denoted using vertices. As a matter of fact, various vertex and edges may utilize certain colors based on the channels that correspond to the SU daily operations, additionally, upon arousal of any conflicts with nearby vertices (e.g., nearby SUs).

Particular collection of estimated calculations is utilized in finding the correct coloring and labeling method which increases the utility rate. The definition of the utility function is based on the throughput values, profit, transmission power, and the interference. Such a technique is upheld through taking into account various proportional equity under consideration. It thus structures the local bargaining power of the graph in which each set of vertex is exhibited by a separate group of the SUs fitting in a similar set of poverty line. Thus the process of coloring specifically makes sure that a separate group acquires a minimum spectrum allocation and therefore makes sure that there is proportional equivalency. In the long run, the issue would have been handled based on a centralized and the distributed fashion systems.

According to the simulation outcomes it is shown that both centralized and distributed solutions comprise of same execution process and cooperative solutions outfit the aggressive solution and nearly get to a global optimum point. In order to contest with the intricacy of the color based spectrum allotment, various parallel computations should be introduced. Despite the fact that the graph coloring can be traced within the distribution fashion channels, there is still need of cooperation amid the clients, therefore, much data exchange is required. With an increase in communication valence an overhead transmission can offset the reimbursement of the decentralized functioning. Additionally, the queuing theory is also analyzed in this paper. The theory mainly involves that study of how effectively the transmission of the CR networks can be handled. Consequently, a queue length centered scheduled mechanism was introduced in order to recognize effective temporal and spatial spectrum gaps used for the retransmission, thus leading to a huge (QoS) quality of service gain for the SUs.

The cost based methodology does not straightforwardly handle the task issue. Additionally it doesn't calculate the real task yet it figures out the value that manages the spectrum task. Specialist co-ops choose the costs for various groups and let clients pick which cost is adequate, thus deciding how clients are doled out.

Somewhat, the cost based methodology settles the range task in circulated mold. A significant number of the evaluating procedures are developed utilizing diversion hypothesis, where each system substance is considered as a player. Every player has his very own target. Administrators need to move the asset for as high a cost as could reasonably be expected, while clients need to get however much asset as could be expected at a given cost. The thought is to think of a legitimate evaluating plan that meets everybody's goal. Contingent upon how the system is demonstrated, the association could be between clients, among clients and administrators, or between various administrators. A standout among the most far reaching ponders around there is done in this paper, where SUs and the system administrators are demonstrated as players in an amusement, each with its own utility capacity that is characterized as a component of cost.

Contingent upon the economic situation and the level of collaboration between various system administrators, an estimating procedure is figured that will put the amusement in harmony and the channel task is therefore decided. An amusement hypothesis is likewise utilized and yet detailed in an unexpected way. Each SU communicates his divert inclination as far as throughput while each channel indicates the sort of SU it lean towards as far as transmission control. The thought is to locate a coordinating channel for each SU dependent on their inclinations. No-lament learning approach is proposed in where both agreeable and non-helpful situations are considered. It merits referencing that valuing system practically speaking is a complex multifaceted issue. Financial elements like impetus to utilize the administration, learning of the clients, value flexibility and request work, social components like social welfare and social decency, and specialized elements like multifaceted nature, estimating interim, and charging system should all be considered. Be that as it may, the valuing plans examined here normally think about just a single or two of these issues.

Various types of algorithms are not compatible with the 3 segments of linear programming techniques which increase the SUs spectrum effectiveness, with a neural networking system which creates interference thus keeping away from spectrum assignment within the ultra wide band CR networking system; the SUs are permitted by the adhoc routing network protocol which ensures switching to dissimilar paths and the spectrum bands to ensure maximum performance, while the Markov process also regulates the admission of the SUs without weakening the Pus and the QoS. A huge number of techniques stated above undergo 3 major flaws. Firstly, they don't scale up very well. When the network comprises a bounded responding time limit, while the demand is very high, such approaches are not in a position of calculating the correct answer based on the specified limit. Secondly, they are also associated with the objective function. In an event when the objective is transitioned or rather the presumed utility function is not correct, such a technique will not be successful. Ultimately, client movement and other traffic related types are thus not paid attention or presumed in this approach. Due to the introduction of the CR idea, uncertainties are further studied for taking advantage of the functionalities for certain applications. Therefore some of the applications are stated below.

10.7   Increased Spectrum Usage

According to Sect. 10.2, it is clearly shown that an emerging spectrum demand initiates perception of the inadequacy of the spectrum. Though, in this case there is a higher supply of the spectrum although it's defectively used. Devices of CR operating systems based on the OSA (opportunistically spectrum access), the use of spectrum will rise up since it will accumulate the gaps through the use of the spectrum gaps. Other effective methods being researched that provide efficient use of the radio resources for the current situation based operations are projected to increase spectrum utility and on the other hand resolving spectrum scarcity issues.

10.7.1   Interoperability

The exchange of data amid vast networking systems or the nodes with the CR being the channel between the two is one likelihood. As a result it has created a channel for brand new wireless applications which highly use the abilities of dissimilar CR network architectures to inter-work, exist, and also seamlessly operate together within the same ecosystem. There is a provision of effective communication links among the wireless communication systems with the aid of interoperability which in turn allows specific units from one or more entities in associating with each other and thus shares thoughts and ideas based on the prearranged technique so that prospected outcomes can be attained. According to IEEE (Institute of Electrical and Electronic Engineers), interoperability is defines as: "the capability of one or two frameworks in sharing data and the utilization of information which has been inter-linked." Interoperability can be acquired in two distinctive methods. They include:

- Syntactic interoperability—this method states that if one or two networks are in a position of sharing data exclusive of any intermediate node or system, then they will be termed to be exhibiting syntactic interoperability. Precise data formats, communication procedures, and others are very essential. Generally, an XML (extensive mark-up language) or rather the SQL may provide interoperability. Finally, the syntactical interoperability is essential for any trial of detailed interoperability.
- Semantic interoperability—basically, with this method, two or more networks are not in a position of linking data of one's own, thus they are in need of extra common system for co-existing. This type involves that ability of automatically analyzing data shared accurately so that to produce outcomes as interoperability, while each side should postpone a common data sharing reference policy.

10.7.2   Exhibition of Interoperability

A potential architecture utilizing such a possibility in the GNU radio cetralized place. It implements a cognitive regulated mechanism with the aid of various algorithms. A different architectural design is known as the plastic project which is a reconfigurable software radio networking system. It is executed by the CTVR (centre for telecommunication value chain research). On the other hand, the CTVR and the CWT (centre for wireless telecommunication) have mutually joined to form a cooperative research aimed at demonstrating and assessing the interoperability of the likelihood with the aid of the CR. Thus the main objective of this project is stated below:

- Exhibition and valuation of interoperability amid two independent advanced CR networks.
- Examination of the aptitude of interoperability of dissimilar architectures in contributing to an ordinary section and the coexistence of interference in a free manner.
- Extending of CR networks and the exploration of possibility of such networks in various scenes.

A different function in the development of the CR networks centered on the interoperability basis is exhibited in a scenario when a general scheme is used for the reconfiguration of the SDR (software defined radios) utilizing a set of wise algorithms instituted. Basically the author makes use of the CR network which is a reference point for the two networks. Therefore the two networks well thought out in this practical are the Wi-Fi and the Wi-Max. Previously analyzed works between the Wi-Fi and the Wi-Max below have been shown. There has been analysis of different techniques of coexistence of the UWB (ultra wide band) using the cognitive radios. In 2005 a high reactive cognitive radio algorithm was anticipated for coexistence among the Wi-Max and the Wi-Fi, it was investigated for the sensing and the interference threshold. There is engagement of the spectrum coordination channeling etiquette protocols for the coexistence amid the two networks which is relatively rationalized with previous outcomes based on the reactive interference evading of the algorithms.

Therefore a NS2 simulation structure is invented in order to analyze the execution process for the representative system events. A different work is thus established in order to create a brand new link within a similar shared frequency within two networks; therefore this will be focused on the semantic interoperability. Such a design constitutes the Wi-Max subnet, the CR, the Wi-Fi subnet, and the CBR which is linked within the two subnets utilizing the CR systems as a unique sharing interface. Therefore, the CR system aids different networks in co-existing [12]. The utilization of unqualified spectrum raises the likelihood of allocating a spectrum opportunity which is used for a common data exchange point within the cognitive network. Therefore, the entire spectrum hole is focused on the semantic interoperability in which the CR is the unique point of reference where data exchange amid the two networking systems occurs, in which it is not in a position of communicating with each other.

10.7.3 Advanced Technologies

Apart from the above stated common use of the CR networks, there are vast applications relating to the CR since there are certain remuneration extracted within the integral technologies such as the OFDM, MIMO, adjustability, self-organization, etc. Various applications include:

Spectrum regulations: As a result of certain benefits acquired from proper spectrum usage, there are other efforts which will in turn reform within dogmatic policies for the allotment of the spectrum.

Link quality: Cognitive radios are quite adjustable during power transmission, intonation schemes, and fault rectification and learning cognition based on previous experience that aids in provision of an excellent quality link.

10.7.4 Additional Wireless Services and Applications

In the upcoming generation, different wireless applications within the CR networking sites are prospected to encounter integral development. Therefore they include: the upcoming generation for the internet service and seamless QoS assurance of various type of multimedia applications, wise distribution channels which makes easy for data sharing and the process enhance the effectiveness of distribution of vehicles. Development within the wireless service utilizing the cognitive radios tends to be wanting during emergency scenarios specifically for the limitation within the electro-magnetic interference in such circumstances.

10.8 Summary

This contribution has presented an analysis of services and application of intelligent spectrum handover by evaluating critical and futuristic features in the field. The services are speculated to lead the foundation of the following network generation either partially or completely. Resultantly, it will be effective for control if band scarcity and any issues arising during spectrum sensing, including the restriction efforts of utilizing assigned bands for handover that is critical in emergent instances. Nevertheless the implementation of intelligent spectrum necessitates through analysis of literature to formulate a remedy which is cost-effective to control the services and challenges discussed in this contribution. Intelligent spectrum handover is fundamental since it ensures that the model of interference causes free operations from cognitive radio services. It is evident from the paper that all nations share an electro-magnetic spectrum to aloofness privileges for its unlimited utility.

References

1. Khan, S., Mitschele-Thiel, A.: Hypernetworks based radio spectrum profiling in cognitive radio networks. EAI Endors. Trans. Cognit. Commun. **1**(2), e5 (2015)
2. Anandakumar, H., Umamaheswari, K.: An efficient optimized handover in cognitive radio networks using cooperative spectrum sensing. Intell. Autom. Soft Comput. 1–8 (2017)
3. Suganya, M., Anandakumar, H.: Handover based spectrum allocation in cognitive radio networks. In: 2013 International Conference on Green Computing, Communication and Conservation of Energy (ICGCE), Chennai, pp. 215–219 (2013)
4. Anandakumar, H., Umamaheswari, K.: Energy efficient network selection using 802.16g based GSM technology. J. Comput. Sci. **10**(5), 745–754 (2014)
5. Kuiper, D., Wenkstern, R.: Agent vision in multi-agent based simulation systems. Auton. Agent. Multi-Agent Syst. **29**(2), 161–191 (2014)
6. Tsuji, H., Tsukamoto, K., Suzuki, K., Nagayama, H.: Development of high-speed mobile radio communication systems using 40 GHz frequency band. Radio Sci. **51**(7), 1220–1233 (2016)
7. Liu, T., Shao, S., Ye, D., Tang, Y., Zhou, J.: Visual cognitive radio. Concurr. Comput. Pract. Exp. **24**(11), 1252–1260 (2011)
8. W Anandakumar, H., Arulmurugan, R., Onn, C.C.: Computational intelligence and sustainable systems. In: EAI/Springer Innovations in Communication and Computing (2019)
9. Wu, Y.: Localization algorithm of energy efficient radio spectrum sensing in cognitive internet of things radio networks. Cogn. Syst. Res. **52**, 21–26 (2018)
10. Su, H., Moh, S.: A directional cognitive-radio-aware MAC protocol for cognitive radio sensor networks. Int. J. Smart Home. **9**(4), 239–250 (2015)
11. Vizziello, A., Amadeo, R., Favalli, L.: Social cognitive cooperation for device to device communications. EAI Endors. Trans. Cognit. Commun. **3**(11), 152557 (2017)
12. Szydelko, M., Dryjanski, M.: 3GPP spectrum access evolution towards 5G. EAI Endors. Trans. Cognit. Commun. **3**(10), 152184 (2017)
13. Grace, D., Zhang, H., Nekovee, M.: Editorial: cognitive communications. IET Commun. **6**(8), 783 (2012)
14. Gurugopinath, S., Muralishankar, R., Shankar, H.: Spectrum sensing for cognitive radios through differential entropy. EAI Endors. Trans. Cognit. Commun. **2**(6), 151147 (2016)
15. Borra, D., Iori, M., Borean, C., Fagnani, F.: A reputation-based distributed district scheduling algorithm for smart grids. EAI Endors. Trans. Cognit. Commun. **1**(2), e3 (2015)
16. Guo, W., Huang, W.: Multicast communications in cognitive radio networks using directional antennas. Wirel. Commun. Mob. Comput. (2012)

Index

Printed in the United States
by Baker & Taylor Publisher Services